recent titles in the *Artech House Microwave Library*,
turn to the back of this book.

Noise in Linear and Nor

For a listing o

Noise in Linear and Nonlinear Circuits

Stephen A. Maas

Library of Congress Cataloging-in-Publication Data
Maas, Stephen A.
Noise in linear and nonlinear circuits/Stephen A. Maas
p. cm.—(Artech House microwave library)
Includes bibliographical references and index.
ISBN 1-58053-849-5 (alk. paper)
1. Electronic circuits—Noise. I. Title. II. Series.

TK7867.5.M33 2005
621.382'24—dc22 2005048030

British Library Cataloguing in Publication Data
Maas, Stephen A.
Noise in linear and nonlinear circuits.—(Artech House microwave library)
1. Electronic circuits—Noise 2. Electronic noise I. Title
621.3'8224

ISBN-10: 1-58053-849-5

Cover design by Igor Valdman

685 Canton Street
Norwood, MA 02062

International Standard Book Number: 1-58053-849-5

10 9 8 7 6 5 4 3 2 1

In memory of
Barry Ross Allen
A colleague and a friend

Contents

Preface

Throughout technological history, noise has been the fundamental problem in communication systems. Indeed, classical communication and information theory deals almost exclusively with the behavior of signals in noise. It was only when extremely low-noise transistors became available, and systems became more complex and distortion-sensitive, that intermodulation distortion (and related phenomena) became equally serious concerns.

In spite of the existence of low-noise transistors that seemed inconceivable only a couple decades ago, noise has again become a matter of great importance. The culprit now is nonlinear noise, a somewhat misnamed set of phenomena in which noise combines with circuit nonlinearities to cause all kinds of trouble. For example, low-frequency noise in transistors, which previously was troublesome only in limited circumstances, modulates the output of oscillators causing phase noise. This, in turn, affects phase- or phase-amplitude-modulated systems in much the same way as high-frequency amplitude noise affects amplitude-modulated systems. The problems engendered by noise never seem to be solved completely; like cockroaches after a nuclear war, they mutate and reappear in other places and other circumstances. An understanding of noise is something we undoubtedly will need, well into the future.

In this book, as in others, I have tried—however successfully—to address both broad theoretical and specific technical aspects of noise theory and circuit optimization. The first six chapters are concerned with the theory of noise in both linear and nonlinear circuits. The theory of linear circuits has existed for some time but has never been collected into a comprehensive book. The first few chapters are an attempt to do that. Nonlinear noise theory similarly has not been well covered in technical books. Previously, anyone interested in this material was forced to resurrect it from a broad range of technical literature. I am optimistic that collecting the information in one place, and explaining it in a straightforward manner, should be valuable to many people who must deal with these matters daily.

The remainder of the book covers applications to various kinds of circuits. Low-noise amplifiers are an obviously important application, and I have tried to cover these in some detail. Noise in mixers, frequency multipliers, and especially oscillators are similarly important considerations in nonlinear circuits. Finally, we examine noise effects in systems as well.

I have quite a few people to thank; in some cases, for their help, and in other cases, for their tolerance. Most important is my wife Julie, who supports me unhesitatingly well beyond the point of anything I could dare to ask for. I am also grateful to the gang at Applied Wave Research for their helpful discussions and ideas, and finally to many colleagues, too many to list, who developed the theory, over the years, that is the subject of this book. Largely, I am restating their work, not describing my own.

Stephen A. Maas
Long Beach, California
August 2005

Chapter 1

Introduction

The earliest problem in electronic communications was to create receivers capable of detecting weak signals. Even with the invention of the vacuum tube, the difficulty of obtaining adequate gain at high frequencies limited the capabilities of receivers. Very soon, however, it became clear that electronic noise was the limiting factor in communications, and efforts were made to minimize both noise and its effects. These efforts continue into the present, even though extremely low-noise receivers can now be created. The goal of this book is to impart a comprehensive understanding of noise, its effect on systems, and methods for its analysis in both linear and nonlinear circuits.

1.1 THE PROBLEM OF NOISE

1.1.1 Noise in Communication Systems

Electronic noise is manifest as a randomly varying voltage or current. The defining characteristic of noise is its random nature. Although they are sometimes loosely called *noise*, deterministic signals, such as spurious signals, interference, and intermodulation distortion, are not noise and are outside the scope of this book. Here, we address only random noise.

In this book, we treat electronic noise as a *stationary random process*. That is, the noise voltage or current at any point in a circuit varies randomly, but with statistical characteristics that do not change with time. Stationarity is an assumption that may not always be warranted; however, in the great majority of cases it is entirely appropriate, and thus will be an underlying assumption throughout.

The noise that affects communication systems can be generated in the electronics or can be received along with the signal. One or the other of these sources may dominate, depending on the frequency and application of the system. In receivers, received noise dominates in the HF region, while internal noise generally dominates at microwave frequencies.

Internal electronic noise in transistors and resistive elements results from a number of physical phenomena. It arises, most fundamentally, from the granularity of charge in electronic devices; that is, current consists of a very large number of individual charge quanta. Therefore, the current resulting from, say, injection of charge into the junction of a diode inevitably is not constant. The time-average value of the injected charge is indeed constant with time, but it still exhibits random variations, which we call *shot noise*.

The most important form of noise we encounter is called *thermal noise*. Thermal noise comes from the random motion of charges in lossy elements and is present in such elements at temperatures above absolute zero. Unlike other forms of noise, thermal noise is predictable from fundamental physical principles. For this reason it is used as a standard for describing other types of noise.

External noise can have various sources. One is *atmospheric noise*, which arises in thunderstorms that occur continuously throughout the world. The level of this noise is greatest at low frequencies, and decreases rapidly with frequency. Another source is *galactic noise*, which is generated by astronomical sources far from the Earth, and is the basis for the science of radio astronomy. Some of these sources, such as the well known radio source Cassiopeia A, are surprisingly strong; the sun is also a powerful noise source. Even thermal radiation from the Earth is a significant source of noise when the receiving antenna, for example, on a satellite, is pointed toward it.

From intuition, one might expect that a signal whose power is well below that of its noise cannot be received reliably. The reality, however, is not so simple. From the earliest days of radio broadcasting, it became clear that, in some cases, a signal-to-noise ratio (SNR) could be improved. One early example is wideband analog frequency modulation (FM). FM exhibits a *threshold effect*; if the signal power is above a threshold, the SNR increases disproportionally with increase in signal power, and, for all practical purposes, the noise quickly disappears. More sophisticated digital spread-spectrum techniques, developed in the second half of the twentieth century, create noise-like signals that can be received reliably even when their spectral density is below that of accompanying noise. These techniques, originally developed for the military, are now used widely in modern commercial systems.

In any case, noise has always been, and continues to be, the fundamental limitation to the capabilities of communication systems. Communication and information theory are almost exclusively formulated around the problem of obtaining reliable communication in a noisy environment.

1.1.2 Phase and Amplitude Noise

A modulated signal in a communication system is a narrowband process. That is, it has a spectrum whose width is small relative to its carrier frequency or to some center frequency within its band. Although the spectrum of the noise may be wider, we generally are concerned only with the noise in approximately the same band as the signal. Thus, the noise also can be viewed as a narrowband process.

A narrowband signal can be viewed as a sinusoid that varies slowly and randomly in phase and amplitude. This case is shown in Figure 1.1. The noisy signal can be viewed as a constant signal phasor, S in Figure 1.1, and a slowly time-varying noise phasor, $N(t)$. The amplitude and phase of the noise phasor varies randomly. The resultant of the two phasors has phase and amplitude components, $\theta(t)$ and $|S + N(t)|$, respectively, both of which are random processes.

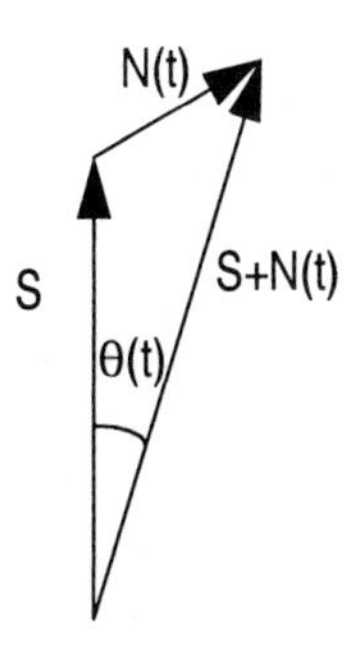

Figure 1.1 In a narrowband noise process, the signal and noise can be viewed as slowly varying phasors. The noisy signal is the phasor sum of the individual components.

The phase component of a noisy signal, appropriately called *phase noise*, is particularly serious in receiver local oscillators. Because a mixer is effectively a phase subtractor, the phase noise of the oscillator signal is transferred in the mixer, degree for degree, to the received signal. If the received signal's phase carries information, the oscillator's phase noise can corrupt it.

Amplitude and phase noise affect communication systems in various ways. Certain systems are especially sensitive to one component or the other. In analog phase and frequency modulated systems, for example, the signal often passes through a limiter, which removes amplitude components entirely, virtually eliminating amplitude noise. The phase noise, however, is not removed, and is turned into baseband noise in the demodulation process. Conversely, amplitude modulated systems are highly sensitive to amplitude noise. Digital communication systems may be sensitive to both. For example, in quadrature amplitude modulated (QAM) signals, information is carried by both the amplitude and phase, so both the amplitude and the phase noise components affect the probability of a detection error.

1.1.3 Linear and Nonlinear Noise

Most theory addresses the problem of noise in linear circuits and systems. We frequently model components in linear systems as black boxes having certain transfer functions and additive noise source at their inputs. The noise level of each source can be defined by a spectral density, often expressed as a noise temperature. In linear circuits, each noisy element is modeled as a noiseless element with a shunt current noise source or, occasionally, a series voltage noise source. The noise at any node in the circuit can be calculated in a straightforward manner; the noise at any point is a superposition of the noise from all the noise sources, with correlations taken into account.

Noise in nonlinear circuits is frequently called *nonlinear noise*, a misnomer, as noise is treated as a *linear* perturbation of the nonlinear circuit, and thus is fundamentally a linear phenomenon. In nonlinear circuits, the situation is more complicated than in linear ones, because the noise process in many cases is modulated by a large-signal control voltage or current somewhere in the circuit. The noise in this case is no longer strictly stationary, because its statistics vary over the period of the large-signal waveform. However, those statistics are the same from cycle to cycle, so the noise can be viewed as periodically stationary. We use the term *cyclostationary* to describe this phenomenon. Cyclostationary noise is important in large-signal, nonlinear electronic components, especially mixers and oscillators.

Phase noise in oscillators is a form of cyclostationary noise. It arises primarily from low-frequency noise in the solid-state device, upconverted to the oscillator frequency through reactive nonlinearities in the transistor.

1.2 NOISE ANALYSIS

Throughout this book, we assume that noise is a small-signal phenomenon. Although large-signal noise is sometimes of interest (e.g., high-level noise interference in communication systems), we do not address such cases here. Because of its small-signal nature, noise can always be treated as a weak signal in a linear or linearized circuit.

In static, linear circuits, we usually wish to know the noise at a particular set of terminals, often the output port. Once the output noise spectral density and correlation characteristics are known, along with the linear characteristics of the circuit, they can be converted to a noise temperature, noise figure, or a set of noise parameters. To determine the output noise voltage or current, we treat each noise source in the circuit as a separate excitation and calculate the resulting output noise, accounting appropriately for the correlations between the sources. This seems at first to be a huge effort, but, in fact, the computational cost of the noise analysis is only slightly greater than that of a conventional linear analysis of the same circuit.

In nonlinear noise analysis, we do much the same thing, although with a couple of complications. First, we encounter both stationary and cyclostationary noise sources, so we must calculate the spectra and correlation properties of those cyclostationary noise sources. The spectra are noise sidebands on each of the harmonics of the large-signal waveform; they are, in general, correlated. Second, we require a formulation of the nonlinear circuit that relates the small-signal voltages and currents at the harmonic sidebands. Such a formulation is inherently linear, as it deals with noise signals that are small—therefore, linear—perturbations of the large-signal waveforms. The problem is not unlike mixer conversion analysis, in which the RF voltage is small relative to the local oscillator waveforms, so the mixing element can be treated as a time-varying linear conductance or transconductance. The vehicle for this transformation is called a *conversion matrix*, which follows logically from that linearization. The analysis using the conversion matrices is analogous to the linear, static analysis.

1.3 CIRCUIT OPTIMIZATION

Analysis of a circuit or system is only half the task; the rest is optimization. Once we know how to analyze a circuit, we must determine methods for optimizing its noise performance. In such simple cases as low-noise amplifiers, the task is usually to minimize the noise figure or noise temperature. In narrowband circuits, the conditions for minimizing noise figure are straightforward to determine. In broadband circuits, however, the situation is not so clear and may require considerable trial and error.

In linear two-ports, the noise figure is a function of the source impedance and four parameters: the minimum noise figure of the two-port, a parameter called its *noise resistance*, and the real and imaginary parts of the source impedance that minimizes the noise figure. That impedance, in general, is not a conjugate match to the input impedance of the two-port. It is usually possible to achieve optimum noise matching over a modest bandwidth. As it is seldom possible to achieve an optimum noise match over a broad bandwidth, the noise resistance, which is a measure of the sensitivity of the noise figure to nonoptimum matching, is an important quantity in broadband amplifiers. In such amplifiers, the usual goal is to minimize the maximum noise figure over the band. This is best achieved by a good noise match near the high-frequency end of the band, where the minimum noise figure is highest, and to suffer greater degradation in noise figure at the lower frequencies, where the minimum noise figure is lower and there is more room for error.

The problem of optimizing nonlinear circuits is considerably more complex. Although considerable research on the noise analysis of mixers has been performed, much less work has addressed their optimization. In any case, thanks to the development of high-frequency, low-noise amplifiers, noise in diode mixers is no longer a great concern. The noise figures of active mixers can be quite high, however, if certain important precautions are not observed in the design. For such mixers, noise analysis is most important to verify that the noise figure is reasonable. Noise in diode mixers, when it is a concern, is best minimized simply by achieving low conversion loss.

Because of its critical importance in many types of communication systems, considerable research on phase noise in oscillators has been performed. Because phase noise is largely engendered by upconversion of low-frequency noise, the best way to minimize phase noise is to use a device having low levels of low-frequency noise. It is well known that microwave FETs are much noisier, in this respect, than bipolar devices.

Although phase noise in oscillators has been recognized as an important phenomenon, phase noise in active multipliers has not been widely investigated. It is clear that a noiseless multiplier exacerbates phase noise of an input signal. A frequency multiplier is, after all, a phase multiplier, so multiplying a signal by a factor n results in multiplication of the phase deviation by n as well. The result is an increase in phase noise by a factor of n^2, or $20 \log(n)$ in decibels. Varactor frequency multipliers exhibit this noise increase almost exactly. Active multipliers, especially those using

FETs, may have substantial low-frequency noise, which can generate additional phase noise through much the same process as in oscillators.

It is axiomatic that anything that can be analyzed can, in some way, be optimized. Thus, the key to optimizing such circuits is first to understand and then to analyze the noise mechanisms. In some cases, this process leads to the conclusion that the best empirically derived performance is already close to the optimum. In others, however, it may lead to insights that could improve performance dramatically.

Chapter 2

Noise and Random Processes

Electronic noise is a random process. That is, it is an apparently randomly varying voltage or current, whose value cannot be predicted precisely from instant to instant. Nevertheless, the noise has certain predictable characteristics, which allow us to characterize it in a useful way. In this chapter, we examine the nature of that process and that characterization.

The purpose of this chapter is to introduce a few concepts related to probability and random processes that are essential for an understanding of noise in circuits. It is not intended to be all-encompassing. For a deeper understanding of this subject, the reader is encouraged to examine the references [2.1–2.6].

2.1 RANDOM PROCESSES

A *random process*, perhaps more properly called a *stochastic process*, is one that generates a set of continuous, randomly varying functions of time. That is, the result of an experiment, ζ, is some time function $x(t)$. Thus, we have a set of functions and events, characterized by some $x(t, \zeta)$. In less academic language, we can think of ζ as an observation, while $x(t, \zeta)$ is the randomly varying function of time associated with that observation. We can focus on some particular observation, which we simply call $x(t)$.

Noise voltage or current is a random process. That is, noise is process $v(t)$ or $i(t)$ that varies randomly with time. Although the quantity is random, and its value is not precisely predictable from instant to instant, the noise has certain uniform properties by which it can be characterized. For example, we often know how much power is available from a source of noise; as it happens, the available power is directly related to the noise statistics.

The study of stochastic processes is a broad area of mathematics. However, for our present purposes, we can limit our consideration to a few special cases. Most important among these is a *continuous*, *stationary*, *Gaussian-distributed* random process. We will define these terms in due course.

2.1.1 Probability Distribution and Density Functions

Although we cannot predict the value of a random process $x(t)$ from instant to instant, we can say something about the probability that $x(t)$ is within some range on any instantaneous observation. This probability is specified by a *distribution function,* sometimes called a *cumulative distribution function.* The distribution function, $F_x(X)$, gives the probability that $x(t)$, at any instant, is below some particular value X. [In our case, to be a little more concrete, $x(t)$ can represent either a current, $i(t)$, or a voltage, $v(t)$.] Thus,

$$F_x(X) = P\{x < X\} \tag{2.1}$$

where $P\{\zeta\}$ is the probability of the event ζ. Since $x(t)$ is continuously varying, it is meaningless to speak of the probability of a particular value; any such value occurs only instantaneously. We can, however, specify the probability that x, on any observation, is within some range. The probability that $X_1 < x < X_2$ is simply

$$P\{X_1 < x < X_2\} = F_x(X_2) - F_x(X_1) \tag{2.2}$$

Since $F_x(x)$ is a probability, we must have

$$0 < F_x(x) < 1 \tag{2.3}$$

and

$$\begin{aligned} \lim_{x \to -\infty} F_x(x) &= 0 \\ \lim_{x \to \infty} F_x(x) &= 1 \end{aligned} \tag{2.4}$$

$F_x(x)$ must be continuous; if it were discontinuous, the probability in (2.2) could have a finite value over an infinitesimal range of x. Finally, $F_x(x)$

must increase monotonically with x, because probabilities must always be positive quantities.

The concept of a distribution function is, of course, applicable to discrete random variables, where the probability of a discrete event has meaning. A similar concept, for such continuous random variables as a noise process, is the *probability density function*. It is defined as the derivative of the distribution function:

$$f_x(x) = \frac{d}{dx}F_x(x) \tag{2.5}$$

The probability of the voltage, on any observation, being in a region between X_1 and X_2 is

$$P\{X_1 < x < X_2\} = \int_{X_1}^{X_2} f_x(x)dx = F_x(X_2) - F_x(X_1) \tag{2.6}$$

The probability density function can be viewed very loosely as a measure of the likelihood of the voltage value at any instant. Indeed, the probability of finding x within a small increment Δx around X_1 is

$$P\left\{X_1 - \frac{\Delta x}{2} < x < X_1 + \frac{\Delta x}{2}\right\} \approx f_x(X_1)\Delta x \tag{2.7}$$

From the foregoing, it should be clear that

$$\int_{-\infty}^{\infty} f_x(x)dx = 1 \tag{2.8}$$

Several probability density functions occur frequently in the study of noise. Perhaps the most important is the *normal*, or *Gaussian* distribution, whose density function is given by

$$f(x) = \frac{1}{\sigma\sqrt{2\pi}}\exp\left(\frac{-(x - X_0)^2}{2\sigma^2}\right) \tag{2.9}$$

Equation (2.9) is the familiar bell shaped curve centered on X_0. The quantity σ is a parameter of the function, which determines its shape; when $|x-X_0| = \sigma$, $f(x)$ decreases to approximately 60% of its peak value. Unfortunately, (2.9) cannot be integrated algebraically; it must be integrated numerically. The integral of a Gaussian curve is called the *error function*, erf(x). It is defined, in normalized form, as

$$\text{erf}(x) = \frac{1}{\sqrt{2\pi}} \int_{-\infty}^{x} \exp\left(\frac{-\zeta^2}{2}\right) d\zeta \tag{2.10}$$

and is tabulated in many handbooks.

Most of the noise processes we study are well modeled as Gaussian-distributed processes. There is a good reason for this: the *central-limit theorem*. The central-limit theorem states that the sum of a large number of independent random variables, each having some arbitrary probability density, tends toward a Gaussian. Noise processes are, in fact, the sum of a large number of small, noise generating events; thus, it is no surprise that the combined effect is inevitably a Gaussian random process.

Another distribution of interest is the *uniform distribution*. It has the density function

$$f_x(x) = \begin{cases} \dfrac{1}{X_2 - X_1} & X_1 < x < X_2 \\ 0 & \text{Otherwise} \end{cases} \tag{2.11}$$

Quantization noise, the noise consisting of leftovers after a sampled waveform is rounded to discrete values, is an example of a uniformly distributed noise process.

Finally, we sometime encounter *Poisson noise*. A Poisson process, sometimes called a *random arrival* process, involves pulses of current or voltage that arrive at random intervals. The probability of the occurrence of k pulses in an interval, τ, is

$$P(k) = \frac{(\lambda\tau)^k}{k!} \exp(-\lambda\tau) \tag{2.12}$$

where λ is the average number of pulses per second. The noise process $n(t)$ consists of the sum of a large number of delayed pulse functions $h(t)$,

$$s(t) = \sum_i h(t - t_i) \tag{2.13}$$

Shot noise, which we discuss in Section 2.3.2, is an example of a Poisson noise process. It is possible to show that Poisson noise approaches a Gaussian distribution when the noise level is large enough that a large number of pulses occur within the length of $h(t)$. This invariably is the case in solid-state devices, so the noise can be treated, to a good approximation, as Gaussian.

2.1.2 Moments of Density Functions

Moments of probability density functions are important in noise analysis, as they can be related to important characteristics of the noise, such as its dc and mean-square values. The first moment of a probability density function, also called its *mean* or *expected value*, is

$$E(x) = \int_{-\infty}^{\infty} x f_x(x) dx \tag{2.14}$$

The mean is the statistical average value of the process.

We expect the statistical mean value to be equal to the time-average value of the noise process. When this is the case, we say that the process is *ergodic*. Thus,

$$E(x) = \int_{-\infty}^{\infty} x f_x(x) dx = \overline{x(t)} = \lim_{T \to \infty} \frac{1}{T} \int_{-T/2}^{T/2} x(t)\, dt \tag{2.15}$$

Ergodicity is an important property, as it means that any pair of samples of a random process, in a statistical sense, are pretty much the same. It also means that statistical quantities can be determined from time-domain measurements. The noise processes we examine in this book are invariably assumed to be ergodic.

The nth moment, in general, is given by

$$m_n = E(x^n) = \int_{-\infty}^{\infty} x^n f_x(x) dx \tag{2.16}$$

and the nth central moment is

$$E((x - X_0)^n) = \int_{-\infty}^{\infty} (x - X_0)^n f_x(x) dx \tag{2.17}$$

where X_0 is the mean value.

The nth moment is a statistical average value of x^n; thus, the first moment is the statistical average value of the process, and the second moment is the statistical average value of x^2, or the *mean-square* value. The second central moment, which is somewhat more useful, is called the *variance*, traditionally noted by the variable σ^2:

$$\sigma^2 = E((x - X_0)^2) = \int_{-\infty}^{\infty} (x - X_0)^2 f_x(x) dx \tag{2.18}$$

The square root of the variance, σ, is called the *standard deviation*.

Since we assume all noise processes to be ergodic, the statistical mean value equals the temporal mean, which is simply the dc component of the noise. Many of the noise processes we consider later in the book have no dc component, so the mean is zero. Similarly, the variance equals the temporal mean-square value of the noise. Dividing mean-square voltage or multiplying mean-square current by resistance provides the power in the process.

Throughout the book, we indicate a mean or average by an overbar. Thus, $\overline{i(t)}$ is the average or dc component of a noise current, and $\overline{i^2(t)}$ is the mean-square value.

2.1.3 Correlation

We shall see, in Chapter 5, that a noisy linear two-port can be modeled as a noiseless two-port and a pair of noise-current sources in parallel with the input and output ports. In principle, these noise currents can be measured by short-circuiting both ports and measuring the currents. Usually, both of those currents depend in some way on the same noise sources in the circuit, so the noise of any internal source becomes a component of the noise in *both* port-current sources. As such, the port-current sources are not completely independent,[1] as they both depend, to some degree, on the same internal sources. We would not expect the port currents to be identical, however, since any internal source might affect the noise at one of the ports more than the others.

This partial dependence is expressed as a *correlation*. The cross correlation function of two real random processes, $x(t_1)$ and $y(t_2)$, which we call $R_{xy}(t_1, t_2)$ is

$$R_{xy}(t_1, t_2) = E(x(t_1) \cdot y(t_2)) = \int_x \int_y x(t_1) y(t_2) f_{xy}(x, y) dx dy \tag{2.19}$$

In (2.19) we assumed the processes to be *stationary*; that is, their statistics are not functions of time. Therefore, we have not included time in the joint probability density function $f_{xy}(x, y)$ and need to integrate only over the ranges of the variables x and y. (In fact, we have implicitly made that assumption throughout this chapter.) Because the processes are stationary, the correlation could have been written $R_{xy}(\tau)$, where $\tau = t_2 - t_1$, a function only of the time *difference*, τ. With these provisions, we express the cross correlation in time as

$$R_{xy}(\tau) = \overline{x(t) \cdot y(t-\tau)} = \lim_{T \to \infty} \frac{1}{T} \int_{-T/2}^{T/2} x(t) y(t-\tau) dt \tag{2.20}$$

and if we further assume the processes to be ergodic, the temporal correlation given in (2.20) is identical to the statistical correlation given in (2.19).

One can also define a correlation of the process with itself. That is,

$$R_x(\tau) = \lim_{T \to \infty} \frac{1}{T} \int_{-T/2}^{T/2} x(t) x(t-\tau) dt \tag{2.21}$$

1. We use this term rather loosely at present; later, we will be more precise.

This autocorrelation function is extremely important, as it is possible to show that the power spectral density function (which we loosely call the *power spectrum* or simply *spectrum*) of a process, $S_x(\omega)$, is simply the Fourier transform of the autocorrelation function:[2]

$$S_x(\omega) = \int_{-\infty}^{\infty} R_x(\tau)\exp(-j\omega\tau)d\tau \tag{2.22}$$

$S_x(\omega)$ is necessarily a real quantity; it is the power per unit bandwidth of the process, a continuous spectrum. Although $R_x(\tau)$ is complex in general (for example, it can be applied to a complex envelope function, discussed in Section 2.2.1), we shall be concerned exclusively with real $R_x(\tau)$. From (2.21) it should be clear that $R_x(0)$ is simply the average power of the process.

In any real system, $S_x(\omega)$ is band limited; if it were not, the process would have infinite power and $R_x(\tau)$ would be an impulse function. Frequently, however, we must model the noise process as a flat spectrum over some frequency range of interest. We call such noise *white noise*, after the concept of white light, which is assumed to have a flat spectrum.

Finally, we note that the cross spectral density function is the Fourier transform of the cross correlation function:

$$S_{xy}(\omega) = \int_{-\infty}^{\infty} R_{xy}(\tau)\exp(-j\omega\tau)d\tau \tag{2.23}$$

This quantity is, in general, complex.

2.1.4 Independent, Uncorrelated, and Orthogonal Random Processes

Our concern about correlated noise sources arises from (as we shall see) a headstrong effort to characterize linear circuit noise by black-box models. That is, we know only the noise currents or voltages at the terminals of,

2. RF and microwave engineers are understandably uncomfortable with the characterization of this quantity, which actually has units of volts squared or amps squared per hertz, as a spectral *power*. This is one of the differences between systems engineers and circuit engineers: systems engineers do not worry much about impedance, since it does not affect their real interest, the signal-to-noise *ratio*. If it helps, imagine that this quantity has been scaled to a normalized impedance of 1Ω.

say, a two-port, but we are unaware of the two-port's internal structure. In fact, the noise at those terminals may arise, in part, from the same sources; it is the sum of contributions from a number of internal sources that we may know nothing about. The various internal noise sources generally contribute in different amounts to the noise at the terminals; some sources may not contribute at all to the noise at one port or another, while others may dominate.

In this case, it is inevitable that the noise current or voltage at one port is somewhat the same as the noise at the other port, but it may not be identical. "Somewhat the same" is an imprecise way of saying that the noise sources are *correlated*. In other cases, it may happen that the noise at the terminals comes from entirely separate internal noise sources that have nothing whatsoever to do with each other. Then, the sources are *independent*, which implies that they are uncorrelated. In either case, we must find some way to specify the degree of correlation. Finally, we can take a hint from the theory of orthogonal functions, and note that in some cases the noise processes are *orthogonal*.

A pair of independent random processes has the property

$$R_{xy}(\tau) = \eta_x \eta_y \tag{2.24}$$

where η_x, η_y are the mean values of the processes $x(t)$ and $y(t)$, respectively. If the processes are zero-mean, $R_{xy}(\tau) = 0$; when $R_{xy}(\tau) = 0$, the process is called *orthogonal*. Thus, independent, zero-mean random processes are orthogonal. (The RF and microwave noise sources we examine in this book have no dc components, so most of the noise processes we examine will be zero-mean. Thus, throughout the remainder of this book, we shall assume that independent processes are uncorrelated and orthogonal, unless we explicitly state otherwise.)

It is easy to show that, for uncorrelated zero-mean processes,

$$\begin{aligned} E\{(x+y)^2\} &= E\{x^2 + 2xy + y^2\} = E\{x^2\} + 2E\{xy\} + E\{y^2\} \\ &= E\{x^2\} + E\{y^2\} \end{aligned} \tag{2.25}$$

because, from (2.19), $E\{xy\} = 0$. Thus, the variance of the sum of two such processes equals the sum of their variances. From ergodicity, the variances are the temporal powers, so the processes combine powerwise. This principle is fundamental to the analysis of noise in high-frequency circuits.

2.2 NARROWBAND RANDOM PROCESSES

2.2.1 Representation of Narrowband Random Processes

We are generally concerned with noise in linear systems. Even in nonlinear systems, noise is usually many orders of magnitude smaller than the signals that the system handles, so the noise can be treated as a linear perturbation of the large signals, and therefore can be handled by linear techniques. Thus, the concept of noise in a linear system is very general. One can, of course, imagine situations where the noise is not small, for example, broadband noise used as a jammer in military systems. Fortunately for both the author and the reader, such cases are well outside of the scope of this book.

By virtue of the superposition principle, we can treat each part of that noise spectrum separately, later combining all the parts that we care about. At any particular time, we need concern ourselves only with the effect of the system on a narrow part of that spectrum.

A narrowband noise process,[3] or other random signal (e.g., a modulated carrier), can be expressed as

$$n(t) = a(t)\cos(\omega t) - b(t)\sin(\omega t) \tag{2.26}$$

where $n(t)$ is the noise voltage or current. The quantities $a(t)$ and $b(t)$ are themselves random processes, called the *in-phase* and *quadrature-phase components* of the process, respectively. ω is a reference frequency. If there exists a dominant sinusoid, such as the carrier frequency of a modulated signal, it is usually best to set ω to that frequency. If no such sinusoid exists, as in a noise spectrum, it is usually most convenient to choose the center frequency of that spectrum.

This function can be expressed in a number of other ways, similar to an ordinary phasor:

$$n(t) = A(t)\cos(\omega t + \phi(t)) \tag{2.27}$$

where

$$A(t) = \sqrt{a^2(t) + b^2(t)} \tag{2.28}$$

---

3. In fact, it need not be all *that* narrowband. In linear circuits, it must be sufficiently band-limited that the spectrum not extend below zero or (God forbid!) to infinity. In nonlinear circuits, it is necessary only that noise sidebands, created by modulating a noise source, do not interfere with each other.

$$\phi(t) = \operatorname{atan}\left(\frac{b(t)}{a(t)}\right) \tag{2.29}$$

The *atan* function must be evaluated in such a way as to return $\phi(t)$ in the proper quadrant. The noise process can also be expressed as

$$n(t) = Re\{A_e(t)\exp(j\omega t)\} \tag{2.30}$$

where $A_e(t)$ is a complex function of time, called the *complex envelope* of $n(t)$. Like any other complex quantity, $A_e(t)$ can be written

$$A_e(t) = A(t)\exp(\phi(t)) \tag{2.31}$$

so

$$\begin{aligned} |A_e(t)| &= A(t) \\ \angle A_e(t) &= \phi(t) \end{aligned} \tag{2.32}$$

The processes $a(t)$ and $b(t)$ vary on a time scale of the inverse of the noise process' bandwidth. Since that bandwidth is generally much less than ω, these functions are much more slowly varying than the sinusoids. A good, intuitive view of the signal is therefore a *quasisinusoid* at frequency ω that drifts relatively slowly and randomly in amplitude and phase.

This formulation represents a signal having a finite bandwidth (e.g., a modulated carrier). To represent noise at a particular frequency we must imagine that it also has some finite, if small, bandwidth. We therefore view the noise as having an incremental or extremely narrow bandwidth (say, 1 Hz) centered on the frequency ω. Then, each sample of the noise is a quasisinusoid whose amplitude depends on the power in that bandwidth and whose phase is a slowly varying random quantity.

2.2.2 Narrowband Noise in the Frequency Domain

In the following chapters of this book, we shall be concerned primarily with a frequency-domain characterization of noise. Usually we consider noise bandwidths that are relatively narrow, narrow enough for their noise spectral densities to be considered flat. Because the noise is in linear circuits, or linear perturbations of nonlinear circuits, we can view noise in a wider spectrum as a superposition of contiguous, narrowband noise processes.

Because a random process is not periodic or limited in time, it cannot be treated directly by a Fourier series or transform. It can, however, be treated by Fourier methods in the following manner. We begin by considering a single sample of the noise over a period of time, T. We can then take the Fourier transform of this sample, as[4]

$$i(\omega, T) = \int_{-T/2}^{T/2} i(t)\exp(-j\omega t)dt \quad (2.33)$$

The power spectrum can be obtained as the limit, as $T \to \infty$, of the average of a large number of such spectral components:

$$S_i(\omega) = \lim_{T \to \infty} \frac{\overline{|i(\omega, T)|^2}}{T} \quad (2.34)$$

For this reason, we traditionally express the noise power spectral density, in the frequency domain, as a mean-square quantity. We can define cross-spectral densities similarly. For example, the frequency-domain cross-spectral density of two processes i and v is

$$S_{iv}(\omega) = \lim_{T \to \infty} \frac{\overline{i(\omega, T)v^*(\omega, T)}}{T} \quad (2.35)$$

which relates directly to correlation in the frequency domain.

2.2.3 Correlation in the Frequency Domain

Continuing with our view of narrowband noise as a quasisinusoid, we can define the frequency-domain correlation between two processes. Following (2.23), the correlation between two noise-current processes having finite but narrow bandwidth centered at the frequency ω is

4. A note on notation. Although frequency-domain voltages and currents are traditionally expressed by upper-case variables, it is traditional to use lower-case variables for noise quantities. Occasionally this risks some confusion with time-domain quantities. In most cases, however, we shall show explicit time functionality [e.g., $i(t)$] for time-domain quantities, and indicate frequency dependence when necessary.

$$C_{i;1,2} = \overline{i_1(\omega)\, i_2^*(\omega)} \tag{2.36}$$

which is, in general, complex. The overbar indicates an average over a large ensemble of long samples of the noise. Of course, the quantity $C_{i;1,1}$ is simply the mean-square value of the noise current:

$$C_{i;1,1} = \overline{i_1(\omega)\, i_1^*(\omega)} = \overline{|i_1(\omega)|^2} \tag{2.37}$$

The *correlation coefficient* between two noise processes is simply a normalized correlation,

$$\Gamma_{12}(\omega) = \frac{\overline{i_1(\omega)\, i_2^*(\omega)}}{\sqrt{\overline{|i_1(\omega)|^2}\,\overline{|i_2(\omega)|^2}}} \tag{2.38}$$

which is zero for independent processes.

We have to be a little careful with this quasisinusoidal representation, because $|i(\omega)|$ is an RMS quantity, not a sinusoidal magnitude. Thus, the power in the process is $\overline{|i(\omega)|^2}R$, not $\overline{|i(\omega)|^2}R/2$, where R is the resistance in which the current exists.

We have used noise current as an example, but analogous representations can be written for noise voltage. In practice, noise in RF and microwave circuits is most frequently expressed as a current, because the physical processes often are most naturally modeled as currents. Even when the noise could be expressed as either a current or voltage, current is preferred, as a current representation is more amenable to circuit analysis by computer.

In this section, we have dealt primarily with spectral quantities as functions of radian frequency, ω. This is consistent with most mathematical and system-theory texts. In circuit analysis, however, we customarily deal with temporal frequency, $f = \omega/2\pi$. Thus, we sometimes must convert between the radian-frequency quantity $X_\omega(\omega)$ and temporal $X_f(f)$. From ordinary analysis, we obtain

$$X_f(f) = 2\pi X_\omega(2\pi f) \tag{2.39}$$

2.2.4 Correlation Matrix

In multiport or multiterminal networks, we express the noise magnitudes and correlations as a *noise-correlation matrix*. In the case of currents, it is

$$\mathbf{C}_i = \begin{bmatrix} \overline{i_1 i_1^*} & \overline{i_1 i_2^*} & \overline{i_1 i_3^*} & \dots \\ \overline{i_2 i_1^*} & \overline{i_2 i_2^*} & \overline{i_2 i_3^*} & \dots \\ \overline{i_3 i_1^*} & \overline{i_3 i_2^*} & \overline{i_3 i_3^*} & \dots \\ \dots & \dots & \dots & \dots \end{bmatrix} \tag{2.40}$$

The elements of the matrix are the complex correlations of the currents at each port or terminal. For simplicity, we leave out the frequency term, as it complicates the expression and adds no real information. Occasionally, the off-diagonal elements of a noise-correlation matrix are given as correlation coefficients, while the on-diagonal elements are represented as correlations.

Throughout the book, we assume that the term $C_{i;j,k} = \overline{i_j i_k^*}$. Occasionally, in other literature, one finds $C_{i;j,k} = \overline{i_j^* i_k}$, which is the complex conjugate of the previous expression. Both forms are used in the technical literature, so the reader should be careful of the form used in any particular case. Especially, one should be careful to verify the forms before copying a correlation matrix from a source of transistor data into a circuit-analysis program.

2.2.5 Why Must We Be Concerned with Correlations?

Throughout a study of linear circuit theory, one never has any occasion to consider correlations. All signals are deterministic and can be combined algebraically. That is, if we create a signal $i_3(t)$ as the sum of two others, $i_3(t)$ and $i_2(t)$, we simply add the functional forms of the current waveforms,

$$i_3(t) = a i_1(t) + b i_2(t) \tag{2.41}$$

where a and b are constants, which the author has introduced to make life a little more interesting. If we were to add frequency-domain phasors in the same manner, we would simply combine them vectorially.

We cannot do this with random processes because we do not know the functional form of the currents. We still have

$$i_3 = ai_1 + bi_2 \tag{2.42}$$

but that is about all we can say. Fortunately, however, we do not care much about i_3, either its time-domain or frequency-domain form. The quantity that we really want to know is its *power*, which requires its mean-square value. Then,

$$\begin{aligned}\overline{|i_3|^2} &= \overline{i_3 i_3^*} \\ &= \overline{(ai_1 + bi_2)(ai_1 + bi_2)^*} \\ &= a^2\overline{i_1 i_1^*} + b^2\overline{i_2 i_2^*} + ab\overline{i_1 i_2^*} + ab\overline{i_1^* i_2} \\ &= a^2\overline{|i_1|^2} + b^2\overline{|i_2|^2} + 2abRe\{\overline{i_1 i_2^*}\}\end{aligned} \tag{2.43}$$

Thus, to determine the power, we must know the correlation between i_1 and i_2. This is provided conveniently by the i_1, i_2 correlation matrix, as well as the mean-square values of both i_1 and i_2.

2.3 PHYSICAL SOURCES OF NOISE

Electrical noise arises in lossy circuit elements and through physical processes in semiconductor devices. It can also arrive unwanted at a receiver's input along with the received signal. In that case, it comes from physical processes in the atmosphere, radiation from the Earth or sun, or even from distant galaxies. Generally, we shall be less interested in the source of the noise, merely resigned to the dismal fact that the noise exists and has to be dealt with. In some cases, however, it is helpful to understand the underlying physics of the noise process, as it affects the manner in which the noise can be characterized. It may also lead to some ideas for minimizing it.

2.3.1 Thermal Noise

In 1906, Einstein predicted that the fluctuating charges in a lossy resistor in thermal equilibrium would result in a noise voltage at its terminals. This phenomenon was first observed in 1928 by Johnson and quantified in the same year by Nyquist. Such *thermal noise*, sometimes called *Johnson noise*, is described by Nyquist's theorem. Nyquist showed that the noise spectral density of the mean-square voltage at the open-circuit terminals of a resistor is

$$S_v(f) = 4KTR \tag{2.44}$$

where R is the resistance, T is absolute temperature in Kelvins, and $S_v(f)$ is the spectral density in V^2/Hz. The available spectral power, $S(f)$, is

$$S(f) = \frac{S_v(f)}{4R} = KT \tag{2.45}$$

This relation is obviously an approximation, as it is constant at all frequencies and thus would have infinite power. The complete expression derived by Nyquist is

$$S_v(f) = 4R\left(\frac{hf}{2} + \frac{hf}{\exp\left(\frac{hf}{KT}\right) - 1}\right) \tag{2.46}$$

where h is Planck's constant and K is Boltzmann's constant, $1.37 \cdot 10^{-23}$ J/K. Planck's constant can be expressed as

$$h = \frac{KT}{f_0} \tag{2.47}$$

where $f_0 \sim 6{,}000$ GHz at room temperature. Substituting (2.47) into (2.46), noting that $f \ll f_0$ in ordinary RF and microwave circuits, and taking the limit as $f \to 0$ gives (2.44). At RF and microwave frequencies, thermal noise can be treated as a white Gaussian noise process.

The situation is, in fact, more general. All lossy substances at temperatures above absolute zero generate electromagnetic radiation, merely by virtue of their temperatures. Such substances are necessarily good absorbers of radiation as well. When an antenna is pointed at an electrically lossy substance, or wires are connected to lossy material, the available noise power can be collected. Remarkably, the available power from such a lossy termination, in the RF and microwave frequency range, is a function solely of its temperature and the bandwidth over which the power is measured. Regardless of the nature of the lossy medium, the available spectral power, $S(f)$, is

$$S(f) = KT \tag{2.48}$$

which is the same as (2.45). The available power, P_a, is

$$P_a = KT\Delta f \tag{2.49}$$

where Δf is the bandwidth.

The available power does not depend on the nature of the resistive load. It depends only on its temperature and the bandwidth of observation, Δf. (Of course, perfect absorptivity or pure resistance is an underlying assumption.) Whether the lossy termination is a resistor or an antenna pointed at an absorber, the available noise power at the resistor's or antenna's terminals is the same.

From (2.44), a resistor can be modeled as a combination of a noiseless resistor and a source, either a series voltage source or a shunt current source. The equivalent circuits are shown in Figure 2.1. The open-circuit RMS voltage at the resistor's terminals is

$$v_{\mathrm{RMS}} = \sqrt{4KT\Delta fR} \tag{2.50}$$

and the short-circuit current is

$$i_{\mathrm{RMS}} = \sqrt{4KT\Delta fG} \tag{2.51}$$

where $G = 1/R$. It is easy to show that terminating either equivalent circuit in Figure 2.1 with a resistor R results in the power given by (2.49) transferred to that load.

Because it is so simple and general, the idea of thermal noise is often used as a way to model a source of white Gaussian noise, even if it is not thermal in nature. Thus, we often speak of the *noise temperature* of a source. The noise temperature is the temperature that would create the same available noise spectral power as the source, if it were a thermal, resistive noise source. We examine thermal noise and the thermal noise model in greater detail in Chapter 3.

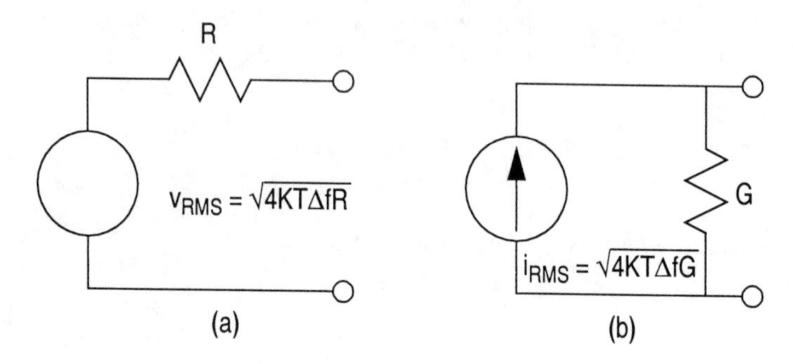

Figure 2.1 Equivalent circuits of a noisy resistor: (a) voltage-source equivalent; (b) current-source equivalent.

It is somewhat less common to define the value of a noise voltage or current source (i.e., the source alone, without the resistor) by a *noise resistance* or *noise conductance*. In this case, the noise resistance or conductance is the value of R or G in (2.50) or (2.51) that provides the correct mean-square value of noise voltage or current; thus,

$$
\begin{aligned}
G &= \frac{\overline{|i|^2}}{4KT_0\Delta f} \\
R &= \frac{\overline{|v|^2}}{4KT_0\Delta f}
\end{aligned}
\tag{2.52}
$$

In this case, the value of temperature must be specified. By convention, embodied in a formal IEEE standard, that temperature, T_0, is 290K.

Because the available power from a matched termination depends only on temperature, matched loads, sometimes heated or cooled, are often used as noise standards. See Section 3.3 for more information on the use of heated and cooled loads as standards in noise measurement.

2.3.2 Shot Noise

Whenever an electron is emitted over a barrier in a semiconductor junction, it generates a pulse of current. The combination of all the pulses created by electrons crossing the barrier necessarily has a fluctuating component, which is a noise process. The mean-square value of the noise is

$$\overline{|i|^2} = 2qI\Delta f \tag{2.53}$$

where q is electron charge, $1.6 \cdot 10^{-19}$ Coul, and I is the junction current, not including the noise itself. I should be interpreted as an unperturbed dc quantity or a deterministic, large-signal current waveform in the device, and $\overline{|i|^2}$ as the mean-square value of a perturbation of that current. Harmonic-balance or time-domain nonlinear analysis can be used to calculate that dc current or current waveform.

In diode junctions, the current consists of both forward and reverse components. Theoretically, both should be included in the noise calculation; however, in modern Schottky-barrier diodes, the reverse current is invariably negligible.

Shot noise has a flat (white-noise) spectrum at frequencies where the transit time of the junction is short relative to the inverse frequency. At higher frequencies, the spectral density decreases with frequency.

2.3.3 Low-Frequency (1/*f*) Noise

All solid-state devices, and even some resistors, exhibit a form of increased noise at low frequencies. The spectrum of this noise varies approximately as $1/f^{\alpha}$, where α is close to 1.0. Frequently, careful measurement shows a spectrum containing regions where α has distinctly different values. This implies that a number of different noise processes exist, each being dominant in its own frequency range. In any case, this noise is often simply called *1/f noise*.

Even today, the physical theory of low-frequency noise is not completely established. Low-frequency noise in solid-state devices seems to be associated with defects and traps in the semiconductor close to interfaces. Devices, such as FETs, in which the electrons travel along an interface, are more strongly affected by low-frequency noise; devices, such as BJTs, in which the electrons cross the interfaces in a perpendicular direction, are less affected.

Low-frequency noise is not as neatly characterized as thermal or shot noise. A common model for such noise has the form

$$\overline{|i|^2} = \frac{k_f I^{\alpha}}{f^{\beta}} \Delta f \tag{2.54}$$

where I is the current in the device and k_f, α, and β are empirical parameters. We assume, in (2.54), that Δf is small relative to f. This type of noise is sometimes called *flicker noise*, a term we use occasionally in later chapters.

2.3.4 Other Noise Sources

Thermal, shot, and low-frequency noise comprise the dominant noise sources in semiconductor devices. Other noise sources may occasionally be significant, or even, in some cases, dominant. A few of these are described in this section.

2.3.4.1 Burst Noise

Burst noise is a type of low-frequency noise occurring in certain types of transistors. It has a $1/f^2$ spectrum with a low-frequency cutoff. One model for the mean-square current in burst noise is

$$\overline{|i|^2} = k_b \frac{I^\gamma}{1 + (f/f_c)^2} \Delta f \tag{2.55}$$

As before, I is the dc or large-signal current in the device, k_b and γ are parameters of the model, and f_c is a cutoff frequency.

2.3.4.2 Hot-Electron Noise and High-Field Diffusion Noise

When electrons are accelerated by high electric fields in semiconductors, they are scattered randomly by interactions with the crystal lattice. Because of their acceleration by the electric field, the free-carrier temperatures of the electrons exceed that of the crystal lattice. The resulting random variations in drift velocity result in current fluctuations, which are manifest as noise.

In FETs, electrons are accelerated to their saturated drift velocity over part of the channel. In this region, the channel is not resistive, and thermal noise concepts are not applicable. Still, the spatial velocity distribution of the electrons is much the same as in a resistive region, because the increase in carrier velocity caused by the electric field is not large relative to the random velocity of the electrons. The increase in carrier temperature, however, increases the noise relative to thermal noise.

High-field diffusion noise in FETs is proportional to the diffusion coefficient of the semiconductor and nearly linearly proportional to channel current. High-field diffusion noise is a dominant noise source in MESFETs and HEMTs.

2.3.4.3 Intervalley Scattering Noise

GaAs and other III-V semiconductors have a satellite valley in their conduction band structures. In high electric fields, some electrons can be scattered into that valley, where their mobility, and hence velocity, decreases abruptly. This change in velocity is manifest as noise.

2.3.4.4 Other Sources

Other sources of noise can be identified; this section, intended only to give the reader some idea of the source of circuit noise, is unavoidably incomplete. In this book, we shall be minimally concerned with the physical sources of noise. We shall be more concerned with methods for characterizing and analyzing noise in RF and microwave circuits.

2.4 CYCLOSTATIONARY NOISE

In the foregoing, we have implicitly assumed noise processes to be stationary. There is one exception that we must consider: *cyclostationary* noise, noise that is stationary in a periodic sense. We noted earlier that many types of noise are functions of a current in a solid-state device; for example, (2.54) and (2.55) show that low-frequency and burst noise, respectively, have a power-law dependence on current. In unpumped circuits, that current is usually a dc quantity, but in mixers or oscillators, it is a large-signal quantity, a function of time. As a result, the noise is modulated by that large-signal current.

Shot noise in a mixer diode is an example of cyclostationary noise. When a mixer diode is pumped by the local oscillator (LO), it rectifies the LO waveform, causing its junction current to approximate a train of sinusoidal pulses. From (2.53), the noise in the junction is

$$\overline{|i|^2}(t) = 2qI_j(t)\Delta f \tag{2.56}$$

where $I_j(t)$ is the (noiseless) large-signal current in the junction. This waveform can be found by straightforward nonlinear analysis, using either time-domain or harmonic-balance methods.

In the static (unmodulated) case, nonoverlapping frequency-domain samples of the noise are uncorrelated. That is, if we were to filter a broadband noise source into two nonoverlapping bands, the two resulting processes would be uncorrelated. However, when a bandlimited noise process is modulated, noise sidebands around each harmonic of the modulating waveform are created, much like sidebands in a mixer. Those sidebands are, in general, correlated, and we must account for their correlation in analysis.

Cyclostationary noise can still be analyzed by linear means, as it can be viewed as a linear perturbation of the large-signal waveform. We examine ways to treat cyclostationary noise in Chapter 6.

References

[2.1] A. Papoulis and U. Pillai, *Probability, Random Variables, and Stochastic Processes*, 4th ed., New York: McGraw-Hill, 2002.

[2.2] M. C. Jeruchim, P. Balaban, and K. S. Shanmugan, *Simulation of Communication Systems: Modeling, Methodology, and Techniques*, New York: Kluwer, 2000.

[2.3] P. J. Fish, *Electronic Noise and Low-Noise Design*, New York: McGraw-Hill, 1994.

[2.4] A. Demir and A. Sangiovanni-Vincentelli, *Analysis and Simulation of Noise in Electronic Circuits and Systems*, Boston: Kluwer, 1998.

[2.5] A. Papoulis, *Circuits and Systems: A Modern Approach*, New York: Holt, Rinehart, and Winston, 1980.

[2.6] J. S. Lee and L. E. Miller, *CDMA Systems Engineering Handbook*, Norwood, MA: Artech House, 1998.

Chapter 3

Noise Figure, Noise Temperature, and the System Noise Model

Although most of this book is concerned with analyzing noise in circuits, we frequently require a system-level, black box characterization. This leads to a two-port noise model in which the two-port is noiseless and has an equivalent thermal noise source at the input. This is, by far, the most common way to characterize noise in components and to predict their performance in systems.

Such a model is necessarily idealized, thus placing restrictive assumptions on the nature of the system in which it is used. Most important is the requirement of matched interfaces; that is, each port is terminated in its standard impedance. Later in this chapter, we examine the implications of this restriction.

3.1 THE SYSTEM NOISE MODEL

3.1.1 Noise Temperature

In Section 2.3.1 we examined thermal noise. We noted that the noise power available from a resistor or any other lossy medium, such as an antenna pointed at an absorber, is simply $KT\Delta f$, where K is Boltzmann's constant and T is its absolute temperature. This *thermal noise* is a white, stationary, zero-mean Gaussian random noise process. Most remarkably, the noise spectral power density, KT, depends only on temperature, and, therefore, the temperature fully defines the available noise spectral power. Noise temperature is equivalent to available spectral power density; the quantities differ only by a proportionality constant. Since we are dealing with linear circuits, we can ignore Dr. Boltzmann and his constant momentarily, and define the noise by its temperature alone. If we need to know the actual power spectral density, we can simply multiply the noise temperature by K.

Because of these characteristics, noise temperature is an ideal way to characterize noise in systems. We can define any noise spectrum as a temperature, T. We need not even feel constrained by the fact that thermal noise is white; if our noise has a nonwhite spectrum, we can simply characterize it by a function of frequency, $T(f)$. The source of the noise need not be thermal; in fact, we generally do not know—or even care—where the noise comes from. All we need to know is its spectrum.

Figure 3.1 illustrates this approach. The figure shows a two-port terminated at its input and output. It is convenient, but not necessary, to imagine that the output port is matched. We imagine that the two-port generates noise having an output temperature T_L, and the input termination has temperature T_s. Finally, we assume, for convenience but not necessity, that the gain and noise spectra are all flat over the bandwidth of interest, Δf. The noise power delivered to the load, P_L, is

$$P_L = K\Delta f(G_t T_s + T_L) \tag{3.1}$$

and its power spectral density is

$$S_L(f) = K(G_t T_s + T_L) \tag{3.2}$$

where G_t is the transducer gain of the two-port. Note that we do not include the noise of the load resistor in this calculation; we are concerned only with the noise delivered to that load, which does not include the noise of the load itself. Because we have assumed that the output port is matched, any noise from that load would simply be dissipated in the two-port's output impedance.

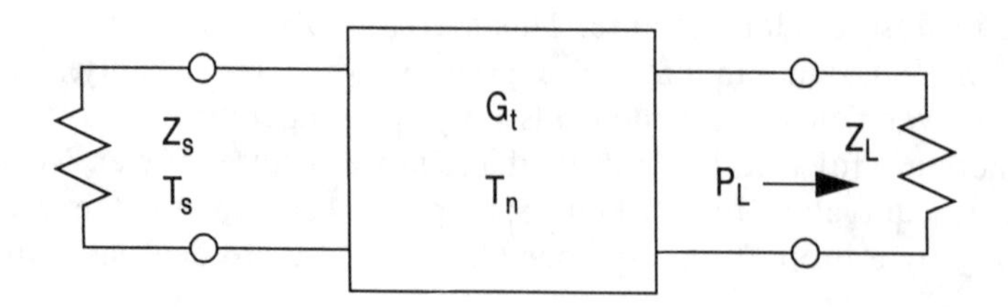

Figure 3.1 A noisy two-port having gain G_t and noise temperature T_n.

Our assumption of a flat gain and noise spectrum is justified by our ability to view the noise over an arbitrarily small frequency range. If we need to extend that frequency range, we can simply view the gain and noise temperatures as functions of frequency, $G_t(f)$ and $T_L(f)$. In this case, (3.1) becomes

$$P_L = \int_{f_1}^{f_2} K(G_t(f)T_s + T_L(f))df \tag{3.3}$$

where f_1 and f_2 are the limits of the frequency range of interest.

This discussion suggests a model for the noise of the two-port. We could simply define the two-port's noise by its output noise temperature, T_L. However, such a model would unnecessarily complicate noise calculations, because they would have to include the two-port's gain in many more cases than necessary. To avoid such complications, we define an equivalent input noise temperature and characterize the two-port's noise in terms of that quantity. The *equivalent input noise temperature*, T_n, is

$$T_n = \frac{T_L}{G_t} \tag{3.4}$$

This model is illustrated in Figure 3.2. The noise of the two-port, T_n, is added, at the input, to any other input noise that might exist. It should be clear, from inspection, that the noise output temperature is

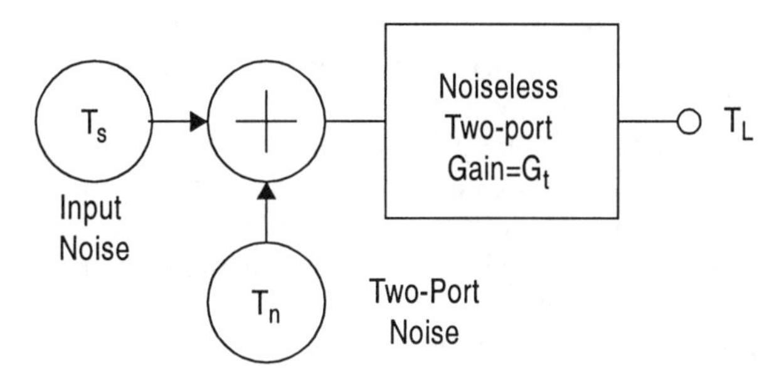

Figure 3.2 Noise model for a two-port. T_n is the equivalent input noise, which is added to any other noise applied to the input. The two-port itself is then treated as a noiseless component.

$$T_L = G_t(T_s + T_n) \tag{3.5}$$

This model is generally applied only to two-ports. It is meaningful only if a single input and output are defined. This requirement is not terribly limiting, because, when multiports are used in systems, it is usually possible to define a specific input and output for each mode of operation. For example, a multiposition microwave switch may have many input ports, but only one—the selected port—operates at a time. When switching is used to allow various system configurations, one simply defines a noise temperature for each configuration.

When several ports of a multiport network are used *simultaneously* as inputs, the noise temperature of the various input ports can be defined more or less arbitrarily. The value of noise temperature at each port does not matter much, as long as the output noise temperature is correct. Specifically, we must have

$$T_{n1}G_{t1} + T_{n2}G_{t2} + T_{n3}G_{t3} + \ldots = T_L \tag{3.6}$$

When we calculate the noise along one path through the multiport, we must include the noise in *all* the ports, as all those noise components contribute to the output noise. Thus, regardless of the division of noise temperatures implied by (3.6), the total output noise, in all cases, is the same. This dilemma is at the heart of the confusing concepts of single-sideband (SSB) and double-sideband (DSB) noise temperatures in mixers, which are examined in Section 3.2.1.

3.1.2 Cascaded Stages

RF and microwave systems often consist of a cascade of two-ports. The model described in Section 3.1 leads logically to an expression for the noise temperature of a cascade of stages. Figure 3.3 shows such a cascade, using the noise model of Figure 3.2. The output noise temperature is found by starting at the input, adding each noise contribution, multiplying by the stage gain, and continuing through the system to the output. Note that, in this case, we are interested only in the noise of the system, not including the noise of the source termination.

The output noise temperature is

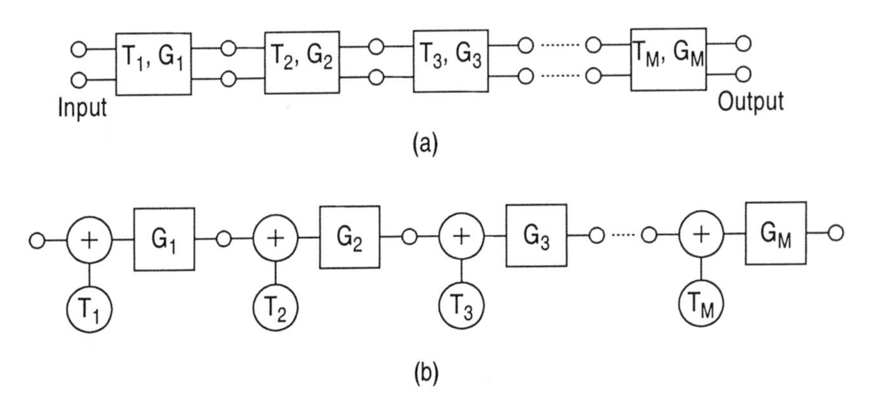

Figure 3.3 Noise model for a cascade of noisy two-ports: (a) cascaded stages; and (b) equivalent noise model.

$$\begin{aligned} T_L &= (((T_1 G_1 + T_2) G_2 + T_3) G_3 + \ldots) \\ &= T_1 G_1 G_2 \ldots G_M + T_2 G_2 G_3 \ldots G_M + \ldots T_M G_M \end{aligned} \tag{3.7}$$

where there are M stages in the cascade. Dividing the output noise temperature by the overall gain of the cascade gives the input noise temperature, T_n,

$$\begin{aligned} T_n &= \frac{T_L}{G_1 G_2 \ldots G_M} \\ &= T_1 + \frac{T_2}{G_1} + \frac{T_3}{G_1 G_2} + \ldots + \frac{T_M}{G_1 G_2 \ldots G_{M-1}} \end{aligned} \tag{3.8}$$

which is known as *Friis' formula*. The contribution of each stage to the input noise temperature is the stage's noise temperature divided by the gain ahead of it. When the gains are greater than unity, the first few stages in the cascade invariably dominate in establishing the noise temperature of the entire cascade.

3.1.3 Noise Figure

An older method of characterizing the noise in a two-port is by its *noise figure*, sometimes called its *noise factor.*[1] The original idea behind noise figure was a measure of the degradation in signal-to-noise ratio as a signal passed through a system. In this sense, the noise figure, F, was defined as

$$F = \frac{(S/N)_i}{(S/N)_o} \tag{3.9}$$

that is, the ratio of input to output signal-to-noise ratios. A moment's reflection should show that this definition is not unique; it depends on the level of the input noise. If the input noise is very small, the F is large, and if it is large, F is small. Clearly, for this concept to be meaningful, the level of the input noise must be defined.

By convention, the standard source noise temperature, T_0, is 290K. 290K is approximately 17°C or 63°F, cool room temperature. This is a convenient value, because an ordinary termination, connected to the input of a two-port, should be close to this temperature. In that case, the noise figure is

$$F = \frac{S/T_0}{G_t S/(G_t(T_0 + T_n))} = \frac{T_0 + T_n}{T_0} = 1 + \frac{T_n}{T_0} \tag{3.10}$$

which is clearly unique. Equation (3.10) can be used to convert between the noise temperature and noise figure of a simple two-port. It also illustrates one reason why we define noise temperature in terms of equivalent input noise instead of output noise: if we had chosen the latter, the term G_t would remain in the equation.

Although (3.10) may be a worthwhile intuitive illustration of noise figure, it is not the precise definition. The precise definition, from an IEEE standard [3.1–3.4], is as follows:

> The noise factor, at a specified input frequency, is defined as the ratio of (1) the total noise power per unit bandwidth at a corresponding output frequency available at the output port when the noise temperature of the input termination is

1. Technically, these terms are identical. Conventional use, however, is for *noise figure* to represent the quantity in decibels, and *noise factor* to represent the simple factor.

standard (290K) to (2) that portion of (1) engendered at the input frequency by the input termination at the *Standard Noise Temperature* 290K.

Mathematically, this is

$$F = \frac{G_t(T_n + T_0)}{G_t \cdot T_0} = \frac{T_n + T_0}{T_0} = 1 + \frac{T_n}{T_0} \tag{3.11}$$

which is precisely the result in (3.10). Although it may be conceptually helpful, there is no real need to define noise figure in terms of signal-to-noise ratio.

The formula for cascaded stages, in terms of noise figure, can be found easily by solving (3.10) for T_n and substituting into (3.8). The result is

$$F = F_1 + \frac{F_2 - 1}{G_1} + \frac{F_3 - 1}{G_1 G_2} + \ldots + \frac{F_M - 1}{G_1 G_2 \ldots G_{M-1}} \tag{3.12}$$

where F_n is the noise figure of the nth stage.

The *noise measure* is defined as the noise figure of an infinite cascade of identical components. It can be found from (3.12) by assuming that F_2 equals the noise measure. Then,

$$F_m = F + \frac{F_m}{G} \tag{3.13}$$

where F_m is the noise measure and F and G are the components' noise figure and gain, respectively. This relation can be solved to produce

$$F_m = \frac{FG}{G - 1} \tag{3.14}$$

The noise measure is a limit to the achievable system noise figure that a component can provide. It is meaningful only when $G > 1$.

Noise figure is traditionally expressed in decibels. Thus,

$$F_{db} = 10\log(F) \tag{3.15}$$

3.1.4 Noise Temperature of an Attenuator

In most cases, it is necessary to measure noise temperatures or to calculate them from the circuit theory that we present in subsequent chapters. An exception is an attenuator, by which we mean any matched, passive, lossy structure. It is possible to calculate the noise temperature of an attenuator simply from its attenuation.

Earlier, we made the point that the noise power available from a matched absorptive termination is $KT\Delta f$, regardless of the nature of the termination. We use this principle to calculate the noise figure of the attenuator. Figure 3.4 illustrates the situation, in which an attenuator has a matched termination on its input, both of which have physical temperature (as opposed to noise temperature) T_p, and we are interested in determining the output noise power. At the output terminals, the combination is indistinguishable from a simple termination, so its output noise temperature must be T_p. Part of this noise comes from the termination, however, and part from the attenuator itself. From Figure 3.2 and our noise model, we have

$$T_L = T_p = G_t(T_p + T_n) \quad (3.16)$$

Solving for T_n gives

$$T_n = T_p(L - 1) \quad (3.17)$$

where the *loss factor* $L = 1 / G_t$. The noise figure can be found by substituting (3.17) into (3.11):

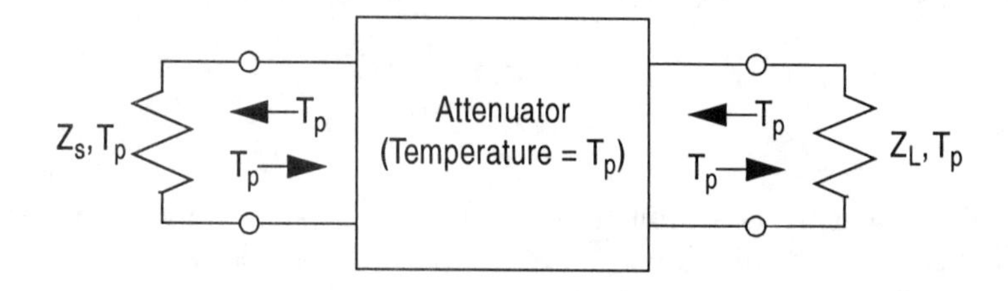

Figure 3.4 When a matched attenuator and its terminations are in thermal equilibrium, the noise power delivered to each port is the same as the power delivered to its termination. This principle is used to derive the attenuator's noise temperature.

$$F = \frac{T_p}{T_0}(L-1)+1 \tag{3.18}$$

and when $T_p = T_0 = 290$, we have the interesting result that $F = L$. Because many systems are operated near this temperature, it is a common practice simply to assume that the noise figure equals the loss.

Before leaving the subject of attenuators, we should look at (3.12) and contemplate the implications of a lossy stage near the input of a low-noise system. Suppose, for example, our system consists of an attenuator followed by a low-noise amplifier. If we assume, for simplicity, that $T_p = T_0$, the noise figure becomes

$$F = L + (F_a - 1)L = LF_a \tag{3.19}$$

where F_a is the amplifier noise figure. We see that the noise figure of the combination is the product of the attenuator loss and the amplifier noise figure, or, in decibels, the sum of the loss and the noise figure. Thus, input loss increases the noise figure of a system one decibel for each decibel of loss. For this reason, great effort is made to minimize input losses in low-noise receivers.

3.1.5 Effect of Idealizations

Throughout this section we have assumed that the components in cascade have a number of ideal characteristics. Primarily, we have assumed that the interfaces are matched; that is, the output impedance of one component and the input impedance of the next one in the cascade are equal to some standard, real, interface impedance, often 50Ω. This requirement is met only approximately in real systems. We should therefore examine the consequences of interface mismatch.

We shall see in Chapter 5 that the noise temperature of a two-port is a function of its source impedance. Thus, we have tacitly assumed that the noise temperature of any component in a cascade is defined for this standard impedance; that is, the impedance of the noise sources used to measure the noise temperature are equal to the standard impedance (Section 3.3).

If the output impedance of the previous stage is not equal to the standard, the noise figure of the following stage changes. In general, the change is not predictable, as it requires data (specifically, the four noise parameters of the two-port) that are not generally available to the system designer.

However, with a few reasonable assumptions, we can generate an intuitive view of the effects of interface mismatch on a two-port's noise temperature.

Consider an amplifier whose source impedance, for optimum noise temperature, is our standard value, Z_0. If we were to measure the noise level at its *input* port, we would notice that a noise voltage exists at the port's terminals. Because of this voltage, the amplifier emits noise from its input toward the output port of the previous stage. If the output impedance of the previous stage is also Z_0, none of that noise is reflected into the amplifier. However, if the output of the previous stage is mismatched, noise is reflected into the amplifier, changing its noise temperature. The amount of the increase (or, possibly, decrease) depends on the level of the emitted noise, the reflection coefficient of the previous stage, and the statistical correlation between the emitted noise and that of the amplifier.

Now, imagine that the amplifier has an isolator at its input. Noise emitted from the amplifier's input port is terminated in the isolator. As long as the isolator's port VSWR is low, the amplifier's noise temperature is the same as with a standard source impedance. Now, however, noise is emitted from the isolator's termination, and the termination noise is uncorrelated with the amplifier's noise. Thus, the level of the noise reflected into the amplifier's input is simply $|\Gamma_o|^2 T_p$, where Γ_o is the output reflection coefficient of the previous stage and T_p is the physical temperature of the isolator's termination. This quantity adds directly to the amplifier's noise temperature. Of course, losses in the isolator increase the noise temperature of the isolator-amplifier cascade in the same manner as any other form of input attenuation.

Interface mismatches also affect the gain of the system, and thus indirectly its noise temperature. The gain of a unilateral two-port with mismatched terminations is

$$G_t = \frac{1 - |\Gamma_s|^2}{|1 - \Gamma_i \Gamma_s|^2} G_{t0} \frac{1 - |\Gamma_L|^2}{|1 - \Gamma_o \Gamma_L|^2} \tag{3.20}$$

where Γ_i, Γ_s, Γ_o, and Γ_L, are the input, source, output, and load reflection coefficients, respectively, and G_{t0} is the transducer gain with a source and load of impedance Z_0. Equation (3.20) shows that an interface mismatch can either increase or decrease the gain, depending on the phases of the reflection coefficients. The phases invariably change with frequency, so the gain of a system having mismatched interfaces exhibits ripple over its bandwidth. As a result, its noise temperature may also exhibit some degree of variation, especially if the gain ripple is caused by a mismatched interface at the system's input.

It is important to note that the input port of an amplifier must be mismatched to optimize the noise temperature. This counterintuitive situation, which we shall examine in detail in Chapter 5, might lead one to wonder whether a chain of amplifiers will cascade properly; that is, whether the gain of the cascade equals the products of the individual gains. Equation (3.20) shows that, as long as the output reflection coefficient of the previous stage is zero, all is well. Of course, the mismatched input increases the sensitivity of the gain to the previous stage's imperfections. This problem can be minimized by the use of an isolator at the amplifier's input, or by using quadrature-coupled, balanced amplifiers ([3.5] and Section 7.1.6.3).

A final question arises: if all else fails, and we are left with one or more badly mismatched stages in a cascade, is there any way to salvage Friis' formula? In fact, there is. If the following changes are made, Friis' formula is still valid:

1. The gains are defined as available gain, with the source impedance of each stage equal to the output impedance of the previous stage.
2. The noise figure or noise temperature of each stage is the value obtained with a source impedance equal to the previous stage's output impedance.

Available gain is defined as P_{ao} / P_{ai}, where P_{ao} is the available power from the output port and P_{ai} is the available power at the input port.

Indeed, there exists a view that Friis' formula should always be defined and evaluated in terms of available gain, as this removes the requirement of matched interfaces. Unfortunately, this practice would not be so simple. In real systems, we rarely know the precise source impedance seen by each stage in the cascade, so the noise figure and available gain are unknown. Furthermore, mismatched interfaces introduce other problems, such as gain and group-delay variations over a bandwidth, that are most easily minimized by matching. Defining Friis' formula in such terms may satisfy academic nicety, but has little value in practice.

An astute reader may have noticed that the IEEE noise figure definition in Section 3.1.3 is given in terms of *available* output power, yet throughout this chapter, we have defined gain as transducer gain, P_{do} / P_{ai}, where P_{do} is the power *delivered* to the load. We have used transducer gain for a number of reasons. First, it makes no difference in the definition of noise figure or noise temperature of a single two-port, since the device noise figure of a two-port does not depend on output matching. Indeed, since we ignore the noise of the load, we could use either gain definition in (3.1) to (3.5), as long as T_L is defined appropriately and the noise generated in the load is ignored. Specifically, with the transducer-gain definition, T_L should be defined as noise power delivered to the load; for the available-gain definition, it should be power available from the output port. Second, when output ports are matched, an assumption we have made repeatedly, there is no difference between available gain and transducer gain. Finally, when we speak loosely of *gain* in microwave technology, we invariably mean transducer gain. Thus, viewing Friis' formula in terms of transducer gain is technically correct, under its stated assumptions, and is consistent with the way we define gain and noise figure or temperature in microwave technology.

3.2 MIXER NOISE

Mixer noise is one of the more confusing subjects in system noise analysis. Mixer noise concepts are complicated by the fact that mixers often have at least two responses, the RF and image. Mixers therefore can be viewed as three-port components, in which two of the ports are actually two responses of a single input port at different frequencies. Furthermore, mixers theoretically have an infinite number of responses, but ones other than the RF and image are often, but not always, well filtered. Conversely, in many cases the mixer input includes an image-rejection filter. Such mixers have no significant image response, so they can be treated as simple, single-response two-ports. The single-sideband/double-sideband concepts described in this section do not apply to such mixers.

The lack of a clear definition of mixer noise figure in the IEEE standard is part of the reason for the confusion. The definition of mixer noise figure, such as it is, is distributed across a number of documents [3.1–3.4]. Consequently, a sort of ad hoc definition of noise figure has evolved [3.6]. Fortunately, this informal standard seems to be widely respected. A third concept, based on an interpretation of the standards, also exists, and has somewhat less use. We call it *the IEEE definition*, for lack of a better term.

3.2.1 Mixer Noise Temperature

In examining mixer noise temperature, we should first recall how noise temperature, in general, was defined. We terminate the input and output ports with standard loads (e.g., 50Ω coaxial terminations), and determine the noise temperature at the *output* port. We then divide this temperature by the gain and subtract the temperature of the input-port termination. In effect, we pretend that the two-port is noiseless and that noise measured at the output arises from an additive source at the input. For this reason, the noise temperature is sometimes called *input-referred noise*.

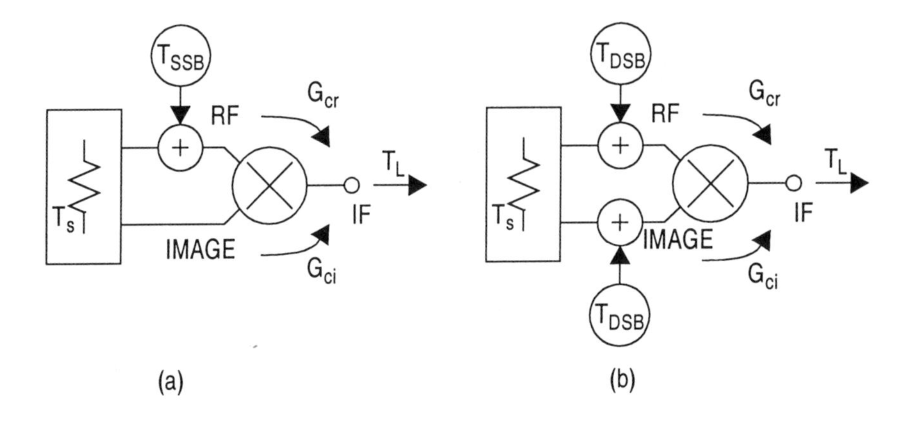

Figure 3.5 Figures illustrating the concept of (a) single-sideband and (b) double-sideband noise temperatures. We treat the RF and image as if they were separate ports, although they actually are different frequency responses at the same physical port.

The same process can be applied to mixers, up to a point. Of course, we can terminate the input port and measure the noise output power at the IF port. A problem arises, however, when we try to refer that quantity to the input. Do we assume that all the noise output power came from the RF port, or that it came from both the RF and image ports? And, if the latter, in what proportions? Answering these questions results in two noise-temperature definitions, which we call *single-sideband* and *double-sideband* noise temperatures. These can be extended to definitions of SSB and DSB noise figure as well.

Figure 3.5 illustrates the two cases. In Figure 3.5(a), we assume that the mixer's noise comes from only the RF port; in Figure 3.5(b), it comes *equally* from both ports. Although this approach seems logical, other ways to assign noise temperatures to the two ports are possible. For example, we could divide the noise between the RF and image unequally, and we would have two different DSB noise temperatures. As long as the division results in the correct output noise power, any such rationale is valid. Fortunately, however, such unnecessarily complicated ideas have not been accepted into general use.

From an inspection of Figure 3.5(a), we find the output noise temperature T_L to be

$$T_L = (T_s + T_{\mathrm{SSB}})G_{cr} + T_s G_{ci} \tag{3.21}$$

where G_{cr} and G_{ci} are the conversion gains from the RF and image ports, respectively, to the IF. T_{SSB} is, of course, the SSB noise temperature and, as before, T_s is the input termination temperature. We solve easily for T_{SSB}:

$$T_{\mathrm{SSB}} = \frac{T_L - T_s(G_{cr} + G_{ci})}{G_{cr}} \tag{3.22}$$

Similarly, from Figure 3.5(b),

$$T_L = (T_s + T_{\mathrm{DSB}})G_{cr} + (T_s + T_{\mathrm{DSB}})G_{ci} \tag{3.23}$$

and from this we obtain

$$T_{\mathrm{DSB}} = \frac{T_L - T_s(G_{cr} + G_{ci})}{G_{cr} + G_{ci}} \tag{3.24}$$

When $G_{cr} = G_{ci}$, a common situation, $T_{\mathrm{SSB}} = 2\,T_{\mathrm{DSB}}$ exactly.

Because diode mixers (as well as other types of passive mixers) are matched lossy components, they are often treated as a kind of three-port attenuator. This is sometimes called the *attenuator noise model* of a mixer. The theoretical underpinnings of this model come from early work on mixers, which concluded that the mixer could be treated as an attenuator having a certain effective diode noise temperature [3.7, 3.8]. As such, we are modeling (as we often do, in the microwave business) a nonthermal noise source as a thermal one. It is possible to show, from a derivation similar to that in Section 3.1.4, that the output temperature of such an attenuator, excluding the termination noise, is

$$T_L = T_p(1 - G_1 - G_2) \tag{3.25}$$

where G_1 and G_2 are the losses through the two paths from the input to the output and T_p, as before, is the attenuator's physical temperature in Kelvins. By analogy, for the mixer, we have

$$T_L = T_d(1 - G_{cr} - G_{ci}) \tag{3.26}$$

where T_d is called the *effective diode noise temperature*. Note that

$$G_{cr} + G_{ci} < 1.0 \tag{3.27}$$

is a fundamental limitation on a diode mixer, so T_L, in (3.26), cannot become negative. $(G_{cr} + G_{ci}) = 1.0$ implies that there is no power dissipation in the diode, and therefore no noise output.

Equation (3.26) can be substituted into (3.23) or (3.24) to obtain an expression for T_{SSB} or T_{DSB}. The most useful case is the common one where $G_{cr} = G_{ci}$. Then,

$$\begin{aligned} T_{\mathrm{SSB}} &= T_d(L - 2) \\ T_{\mathrm{DSB}} &= T_d\frac{(L - 2)}{2} \end{aligned} \tag{3.28}$$

where L, the *conversion loss*, is the inverse of the conversion gain. Of course, if the image response is negligible, the mixer noise temperature T_{mx} is

$$T_{mx} = T_d(L - 1) \tag{3.29}$$

and there is no $T_{\mathrm{SSB}} / T_{DSB}$ duality.

This approach is useful because, in diode mixers having approximately equal RF and image responses, T_d is usually near 350K. This general rule, along with (3.28), can be used to make rough predictions of mixer noise temperatures.

3.2.2 Mixer Noise Figure

Noise figure is not an intrinsically useful quantity, as it must be converted to some measure of noise power, usually a noise temperature, to represent a meaningful physical quantity. When unambiguously defined, noise figure may be valuable as a figure of merit. Easily expressed as a ratio in decibels, noise figure satisfies the RF or microwave engineer's instinctive desire to express all quantities logarithmically.

It should come as no surprise that using a fundamentally two-port quantity as a figure of merit for a three-port component creates an awkward situation. This awkwardness is by itself a good argument for eliminating the use of mixer noise-figure concepts. However, since mixer noise figure is used regularly in industry, and there seems to be no indication of its near-term demise, we need to deal with it.

Dealing with the image-frequency termination noise is the fundamental problem in defining mixer noise figure. The IEEE definition, quoted in [3.1], states,

> For heterodyne systems, [the output noise engendered by the input termination] includes only that noise from the input termination which appears in the output via the principal-frequency transformation of the system, and does not include spurious contributions such as those from an image frequency transformation.

This statement can be interpreted in a couple of different ways. The most common interpretation is that the source-termination noise, at frequencies other than the RF, should not be included in calculating the noise output caused by the source. Although the definition does not say so explicitly, it has always been assumed that termination noise at mixing frequencies other than the RF also should not be included in the calculation of total noise output power. That is, *termination* noise is not *mixer* noise, so the termination noise at the image frequency, or at any other mixing frequency, should not be part of the mixer noise in defining noise figure. This idea is easy to justify; indeed, in defining T_{SSB}, we did not include the termination noise as part of the mixer noise. On the other hand, that noise appears, for all practical purposes, as if it were a mixer noise source. Perhaps it should be included in any figure of merit describing a mixer's noise.

One can make the argument that, as long as the path from noise figure to temperature is well defined, the nature of the noise-figure definition does not really matter. This claim is valid until some poor benighted soul, who fails to understand this admittedly confusing subject, uses the wrong definition in (3.12), Friis' equation in noise-figure form. We examine this question in detail in Section 3.2.4.

Accordingly, we mindlessly apply the above definition and (3.11). (Note that, in any noise-figure definition, $T_s = T_0 = 290\text{K}$.) Ignoring the image termination noise, we find the output noise temperature to be $G_{cr}\,(T_{SSB} + T_0)$ and the noise from the termination alone to be $G_{cr}\,T_0$. The noise figure is the ratio of these quantities,

$$F_{\mathrm{SSB1}} = \frac{T_{\mathrm{SSB}}}{T_0} + 1 \tag{3.30}$$

We call F_{SSB1}, as given in (3.30), the *IEEE definition*. Substituting (3.28) with $L = 1 / G_{cr}$, we have

$$F_{\mathrm{SSB1}} = (L-2)\frac{T_d}{T_0} + 1 \tag{3.31}$$

which reduces to $F_{\mathrm{SSB1}} = L - 1$ when $T_d = T_0$. Previously, we saw that, in an attenuator, $F = L$ when $T_d = T_0$. That is clearly not the case in a diode mixer, when this noise-figure definition is used.[2]

A second option is to include the image noise. In this case, the image noise is treated as a noise source within the mixer, and the output temperature becomes $G_{cr}\,(T_{\mathrm{SSB}} + T_0) + G_{ci}\,T_0$. Assuming the conversion gains to be equal, we obtain

$$F_{\mathrm{SSB2}} = \frac{T_{\mathrm{SSB}}}{T_0} + 2 \tag{3.32}$$

where F_{SSB2} is the noise figure that includes the image noise. Substituting and defining the conversion loss L as before gives

$$F_{\mathrm{SSB2}} = (L-2)\frac{T_d}{T_0} + 2 \tag{3.33}$$

and when $T_d = T_0$, $F_{\mathrm{SSB2}} = L$. This result is hardly a surprise, as the mixer's power-dissipating elements, including the image termination, all have temperature T_0, so it is the same situation as a simple attenuator.

Finally, we determine the DSB noise figure. From Figure 3.5(b), the noise output temperature is $G_{cr}\,(T_{\mathrm{DSB}} + T_0) + G_{ci}\,(T_{\mathrm{DSB}} + T_0)$ while the fraction from the input termination is $(G_{cr} + G_{ci})\,T_0$. The DSB noise figure, F_{DSB}, is

---

2. This result appears to imply that a lossless mixer, presumably with $L = 1$, has a negative noise figure. In fact, one can show that a passive mixer having an image response equal to its RF response must have at least 3-dB conversion loss. Thus, $L \geq 2$ in all such cases, and the minimum noise figure 1.0.

$$F_{\text{DSB}} = \frac{T_{\text{DSB}}}{T_0} + 1 \tag{3.34}$$

which can be expressed in terms of conversion loss as

$$F_{\text{DSB}} = \left(\frac{L}{2} - 1\right)\frac{T_d}{T_0} + 1 \tag{3.35}$$

where, again, with $G_{cr} = G_{ci} = 1 / L$. When $T_d = T_0$, $F_{\text{DSB}} = L / 2$. In this case, the DSB noise figure is half that of F_{SSB2}, just as the DSB noise temperature is half the SSB. In contrast, F_{DSB} and F_{SSB1} approach a 3-dB difference only as $F_{\text{SSB1}} \rightarrow \infty$.

3.2.3 Multiresponse Mixers

This discussion of mixers having only an RF and an image response opens a greater question: how shall we deal with mixers that have even more responses? In general, mixers have small-signal responses at the frequencies

$$f = nf_{\text{LO}} \pm f_{\text{IF}} \tag{3.36}$$

where f_{LO} and f_{IF} are the LO and IF frequencies, respectively, and n is a positive integer. (Here we assume that the IF is the lowest mixing frequency of the set; that is, the mixer is not an upconverter.) In most cases, all but the RF and occasionally image responses are filtered, so other responses are insignificant.

A broadband, doubly balanced diode mixer, however, may have a multioctave—even a decade—bandwidth. When such a mixer is operated near the lower end of its band, the responses associated with several harmonics may be significant. Worse, the source impedance at these high-order mixing frequencies affects both the conversion efficiency and the noise temperature associated with the RF response. As a result, when the mixer is used in a system, which necessarily has limited bandwidth, these responses may be very different from what is calculated or measured in a broadband test set. We examine these cases separately.

3.2.3.1 Analysis

Modern nonlinear circuit simulators include a capability for nonlinear noise analysis. In performing the nonlinear noise analysis of broadband mixers, we face a dilemma: should we include the noise from the terminations at unwanted mixing frequencies? In narrowband mixers, which have substantial RF, LO, and IF filtering, the diode terminations at those frequencies are part of the mixer's structure, and therefore their noise is arguably part of the mixer noise. In broadband, doubly balanced mixers, however, the main source of such noise is the RF termination.[3] For example, imagine a broadband ring mixer that covers 1 to 1,000 MHz. Such mixers are readily available from many commercial sources. We operate this mixer at f_{LO} = 10 MHz and f_{RF} = 11 MHz. All significant LO harmonics and mixing frequencies are clearly within the RF band, and the diode terminations at these frequencies are the RF source impedance. This impedance is clearly extrinsic, and it is not clear whether it should be treated as part of the mixer noise.

Suppose we choose not to include the termination noise in the mixer noise temperature. We then calculate the mixer's SSB noise temperature, T_{SSB}, assuming that the termination impedances are noiseless. Then, when the mixer is used in a system, the contribution to the IF output noise temperature from the terminations, T_L, is

$$T_L = \sum_i G_{ci} T_{si} \tag{3.37}$$

where the summation is over the responses, not including the RF, and T_{si} is the source temperature at each response, and G_{ci} is the conversion gain from the termination at that response to the IF. If the source is a simple 50Ω load, T_{si} is its physical temperature for all i. However, the source may be the output port of a low-noise amplifier, whose noise temperature may vary considerably over the mixing-frequency range. The SSB noise temperature of the terminated mixer, T_{mx}, is

$$T_{mx} = \frac{T_L}{G_{cr}} + T_{SSB} \tag{3.38}$$

---

3. It is possible to show that the output noise, at certain mixing frequencies, originates in the IF output termination itself. We shall ignore the complications that this fact creates, as they do not change the fundamental issue we address in this section.

and this is the quantity used in (3.8) to determine the noise temperature of the system. Unfortunately, when the mixer is used in a system, the source impedances, as well as their noise temperatures, at all the response frequencies may not be the same as in the analysis, and some error is then introduced.

3.2.3.2 Measurement

When a mixer's noise temperature is measured, we face much the same situation described in Section 3.2.3.1. That is, the test set is broadband, so the input port is terminated in a low-VSWR source at many unwanted responses. The output noise, then, is given by (3.37).

In measurement, however, we have an additional problem: the noise source used in the test is also broadband and injects noise into the mixer via the unwanted responses. This noise is downconverted to the IF along with the source noise at the RF frequency, and increases the change in output noise when the hot noise source is switched to the mixer's input. The result is an increase in the Y factor (Section 3.3), which decreases the measured noise temperature. In effect, the measured noise temperature looks better than it should, while the noise at unwanted mixing frequencies should actually make it worse.

This discussion shows that, for valid mixer noise measurements, the noise source must be filtered to prevent noise injection via unwanted responses. Furthermore, the mixer must be terminated in the same manner as in the system in which it will be used; in many cases, this means measuring the mixer and its input filter together, as if they were a single unit.

3.2.3.3 System Application

When a mixer's noise temperature is predicted by a nonlinear circuit simulator, the analysis is performed with an ideal source termination. In use, however, the source termination may be anything but ideal; it is often a filter, which is highly reactive outside its passband. This filter affects not only the unwanted responses that it is designed to eliminate, but, by changing the source impedance at those frequencies, also affects the conversion performance at the *desired* (RF) response. Thus, the simulator's calculation of the mixer's noise temperature, however defined, may be very different from what is observed in an application.

The obvious solution is to analyze, measure, and apply the mixer and its input filter as a unit. If both are implemented as a single monolithic integrated circuit, such an analysis may be practical. However, if the mixer and its input filter are separate units, connected by a length of transmission line, it may be impossible to predict the electrical distance from the filter's output to the mixer's input. Then, the phase of the source reflection coefficient at each response frequency is uncertain. In that case, the only practical solution might be to vary the electrical distance over its range of probable values and to avoid values, in application, that result in poor performance.

3.2.4 Friis' Formula with Mixers

We now examine the question of performing system noise calculations with mixers having a significant image response. As we shall see, the use of a mixer having an image response equal to its RF response generally doubles the system noise temperature. For this reason, in most communication systems, the image response is removed by filtering or, when filtering is impossible, by a so-called *image-rejection mixer* [3.6]. Conversely, certain types of receivers (e.g., radiometers) can use an image response to improve their sensitivity.

To avoid getting lost in a morass of nonintuitive algebra, we shall describe the system analysis by an example. Extension to more general cases is straightforward. Figure 3.6 shows a cascade of two amplifiers and a mixer. As before, the mixer's RF and image responses are illustrated as separate ports. We assume that the RF and image responses are identical, and that the mixer is described by its SSB noise temperature. Furthermore, we assume that the amplifiers' noise temperatures and gains are equal at the RF and image frequencies.

The output noise temperature from the two amplifiers, at the input of the mixer, T_{LA}, is

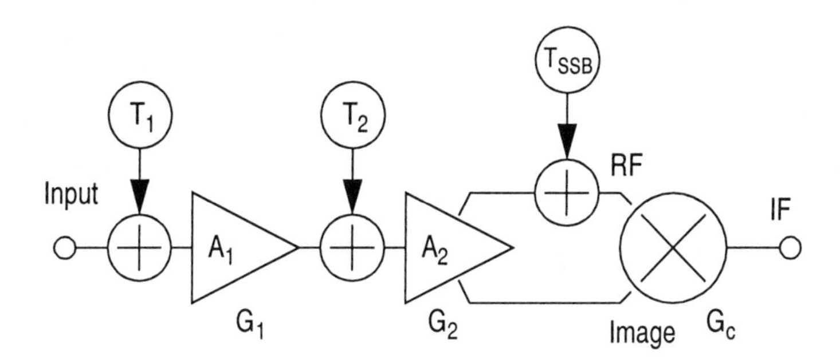

Figure 3.6 A cascade of two noisy amplifiers and a mixer.

$$T_{LA} = T_1 G_1 G_2 + T_2 G_2 \tag{3.39}$$

where the noise temperatures and gains are defined in the figure. The mixer's output noise temperature is

$$T_L = 2T_{LA} G_c + T_{\mathrm{SSB}} G_c \tag{3.40}$$

Substituting (3.39) into (3.40) and dividing by the overall gain, $G_1 G_2 G_c$, gives the SSB system noise temperature:

$$T_{n,\,\mathrm{SSB}} = 2T_1 + 2\frac{T_2}{G_1} + \frac{T_{SSB}}{G_1 G_2} \tag{3.41}$$

From Section 3.2.1, the system's DSB noise temperature is half this quantity:

$$T_{n,\,\mathrm{DSB}} = T_1 + \frac{T_2}{G_1} + \frac{T_{SSB}}{2G_1 G_2} \tag{3.42}$$

If the mixer's DSB noise temperature is given, simply substitute 2 T_{DSB} for T_{SSB} in (3.40) to (3.42). Finally, we can obtain an expression for noise figure by substituting the expressions in Section 3.2.2. The expression we obtain depends upon our choice of noise figure definitions. The second SSB noise figure definition gives

$$\frac{T_{n,\,SSB}}{T_0} + 2 = F_{n,\,SSB2} = 2F_1 + \frac{2(F_2 - 1)}{G_1} + \frac{F_{SSB2} - 2}{G_1 G_2} \tag{3.43}$$

the result is an SSB noise figure that best corresponds to the second noise figure definition in Section 3.2.2.

3.3 NOISE MEASUREMENT

Noise temperature measurements are necessary both to characterize components and to check the results of circuit designs. Fortunately, methods for such measurements are straightforward. They depend on the existence of noise standards, which are frequently cooled or heated resistors. For higher levels of noise, diode noise sources can be used.

All noise measurements derive from a method known as the *Y-factor method*. Because of its importance, we describe it in some detail.

3.3.1 Y-Factor Measurement

In principle, one could simply measure the output noise from a component and, knowing the gain and bandwidth, use (3.1) to calculate the noise temperature. The problem in this approach should be clear after a moment's thought: we generally do not know the gain of the component or the bandwidth of the measuring instrument with sufficient accuracy. A few tenths of one decibel error in these quantities result in the same magnitude of error in the noise temperature. Invariably we wish to measure the noise temperature to greater accuracy.

A better method is one in which the bandwidth and gain do not affect the measurement at all. This can be achieved by comparing the equivalent input noise temperature to terminations having known temperatures. We need at least two such terminations. Traditionally they are called a *cold load* and a *hot load*. Usually, for simplicity, one of these terminations is an ordinary, standard termination at room temperature. The other termination can be heated or cooled to a temperature moderately above or below room temperature. The temperature difference between the hot and cold load is not critical, but for best accuracy, it should be on the same order as the noise temperature to be measured. The noise sources can also be nonthermal noise generators, such as avalanche-diode noise generators, calibrated by comparison to thermal sources.

The test system is shown in Figure 3.7. The device under test (DUT) can be a simple component or a complete receiver. The power meter can be any type of tunable, bandlimited power indicator, but is usually a test receiver with a power detector connected to its IF output. Such a receiver provides measurement at the desired frequency and control of the band-

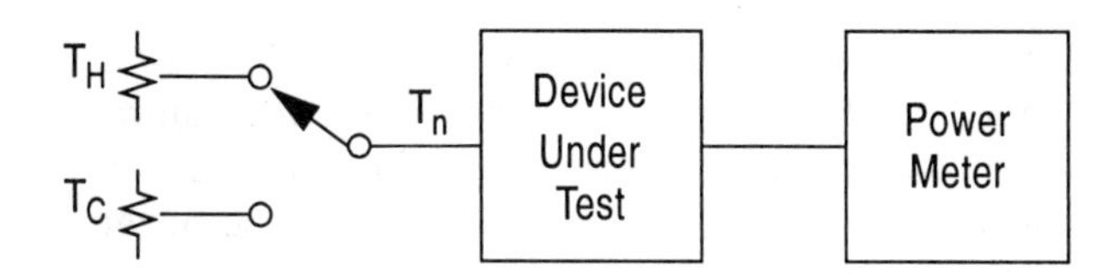

Figure 3.7 Test setup for Y-factor measurements.

width over which the power is measured. We assume that its noise is negligible compared to the noise output power of the DUT. We further assume that the switch at the input is lossless. [In any case, the effect of a noisy power meter or any significant loss in the switch can be removed by the use of (3.8) and (3.17).]

When the switch connects the input to the hot load, whose noise temperature is T_H, the power meter indicates P_{LH}. From (3.1),

$$P_{LH} = K\Delta f G_t(T_H + T_n) \tag{3.44}$$

and when the cold load is connected, the output power, P_{LC}, is

$$P_{LC} = K\Delta f G_t(T_C + T_n) \tag{3.45}$$

Dividing (3.44) by (3.45), we obtain the so-called *Y factor*,

$$Y = \frac{P_{LH}}{P_{LC}} = \frac{T_H + T_n}{T_C + T_n} \tag{3.46}$$

Finally, we solve for T_n:

$$T_n = \frac{T_H - YT_C}{Y - 1} \tag{3.47}$$

which is independent of G_t and Δf.

The Y factor, which is simply a power ratio, can be measured in a number of different ways. One is, of course, to use a conventional power meter or a calibrated detector. Another is to use a precision variable attenuator at the test receiver's output port, followed by a detector and digital meter. In this case, the switch is operated and the attenuator is adjusted until the meter returns to its original value. The Y factor is the change in the attenuation.

In the preceding discussion, we have assumed the power sensor to be noiseless. In practice, we often measure the noise temperature of a two-port whose gain is relatively low and must use a relatively noisy test receiver as the power sensor. In this case, the test receiver's noise becomes a significant part of the measurement. To eliminate its effect, we must measure the receiver's noise temperature and the gain of the DUT; then we use (3.8) to calculate the noise temperature of the DUT alone. Section 3.3.2 shows an elegant way of doing this.

In this analysis we have assumed the hot and cold loads to be ideal; that is, their impedances are equal to the standard impedance of the system, Z_0. Noise temperature, in general, is a function of source impedance, so the noise temperature given by (3.47) is valid only when that source impedance is used.

If the load impedances are different, there is obviously some uncertainty in the noise temperature. Furthermore, the difference in impedances causes the gain of the system to change when the switch is operated. That gain difference affects the Y factor directly.

Note that it is not necessary to assume that the *input* impedance of the DUT is Z_0; the *source* impedance is the important quantity. In fact, unless some form of isolation is used at its input, the input port of a low-noise amplifier generally is significantly mismatched. We shall examine these matters further in Chapters 5 and 7.

3.3.2 Simultaneous Measurement of Gain and Noise Temperature

It is possible to use noise to measure gain. This capability can be quite useful in making automated noise measurements, as it allows for the automatic de-embedding of the noise of the test receiver.

The method is based on a simple principle. The transducer gain of a two-port is

$$G_t = \frac{P_d}{P_a} \tag{3.48}$$

where P_d is the power delivered to the load and P_a is the available power of the source. Since the gain is a linear quantity, it can be expressed as

$$G_t = \frac{\Delta P_d}{\Delta P_a} \tag{3.49}$$

When we switch between the hot and cold load at the input,

$$\Delta P_a = K \Delta f G_t (T_H - T_C) \tag{3.50}$$

and we can measure the resulting change in output temperature, ΔT_L. We assume that the output is matched; the power dissipated in the load is

$$\Delta P_d = K\Delta f G_t(\Delta T_L) \tag{3.51}$$

so

$$G_t = \frac{\Delta T_L}{T_H - T_C} \tag{3.52}$$

Thus, the problem devolves into one of measuring the output temperature of the two-port. The test setup is shown in Figure 3.8. The power sensor in this case is invariably an RF or microwave receiver, which is not noiseless. Its noise temperature is determined by a Y-factor measurement using the hot and cold noise sources T_{H2} and T_{C2} in the figure.

The input switch is connected to T_{C1} and the test receiver is switched between the terminations T_{H2}, T_{C2}, and the output of the two-port, which has noise temperature T_{LC}. The noise output power of the test receiver, usually measured at its IF, is recorded. We call these values P_{H2}, P_{C2}, and P_{LC}, respectively. We then calculate

$$\begin{aligned} \Delta P_1 &= P_{H2} - P_{C2} = k(T_{H2} - T_{C2}) \\ \Delta P_2 &= P_{LC} - P_{H2} = k(T_{LC} - T_{H2}) \end{aligned} \tag{3.53}$$

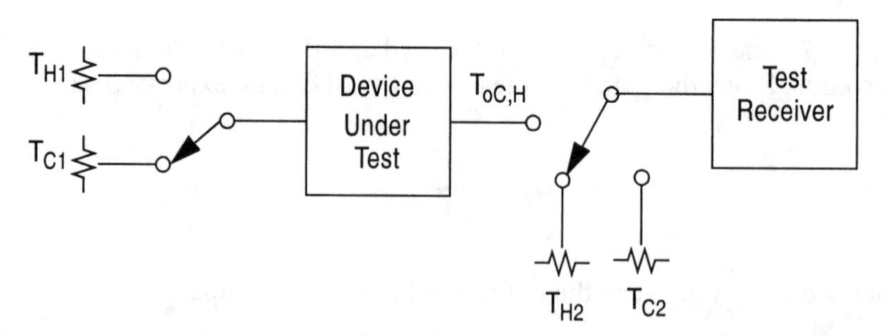

Figure 3.8 Test setup for simultaneous noise and gain measurements. The input switch is used in much the same manner as for Y-factor measurements. The output switch and hot/cold loads calibrate the test receiver so the output temperature of the DUT can be determined.

where T_{LC} is the output temperature of the DUT when T_{C1} is connected to the input, and k is a proportionality constant, which by now should be easily recognizable as $K\Delta f G_r$, where G_r is the gain of the test receiver. (Fortunately, as in previous analyses, this constant disappears, so we need not worry about its precise value.) A little algebra gives

$$T_{LC} = (Y_m + 1)T_{H2} - Y_m T_{C2} \tag{3.54}$$

where $Y_m = \Delta P_2 / \Delta P_1$.

The input switch is then set to T_{H1}, the output to the DUT, and P_{LH} is recorded. P_2 is then recalculated using P_{LH} instead of P_{LC}, and T_{LH} is similarly obtained. The gain is found from (3.52) with $\Delta T_L = T_{LH} - T_{LC}$.

Note that it is not necessary to know the noise temperature of the DUT to make this measurement. The noise temperature at the input of the DUT, however, is easily derived from the measurements, by means of the Y-factor method with $Y = P_{LH} / P_{LC}$. Similarly, the noise temperature of the test receiver can also be determined from the Y factor $Y = P_{H2} / P_{C2}$. Knowing the gain of the DUT and the noise temperature of the receiver, we can use (3.8) to remove the effect of the receiver's noise and obtain the noise temperature of the DUT alone.

Again, it is important to note that we have assumed the output impedance of the DUT to be identical to that of the hot and cold loads T_{H2} and T_{C2}. If this were not the case, the noise temperature of the test receiver, as well as its gain, would change when it was switched from the hot/cold load to the DUT's output port. This assumption is somewhat limiting, as the output VSWR of, for example, a mixer often is 2.0 or higher, while that of the loads is quite low. For this reason, it is a good practice to use an isolator ahead of the test receiver.

3.3.3 Measurement of Reflection Coefficient

Noise measurements can be used to determine the magnitude of a port reflection coefficient. The method is certainly not the first choice for measuring such parameters, as sinusoidal measurements are simpler and more accurate. However, reflection coefficient can be a useful by-product of ordinary noise measurements. Clearly, this technique can be included in the simultaneous measurement of noise and gain described in Section 3.3.2; it is often used in mixer measurements to determine the mixer's IF-output reflection coefficient, an important quantity.

Figure 3.9 shows the test setup. Noise from the noise diode is injected into the port of the device under test, often the output port. Again we use

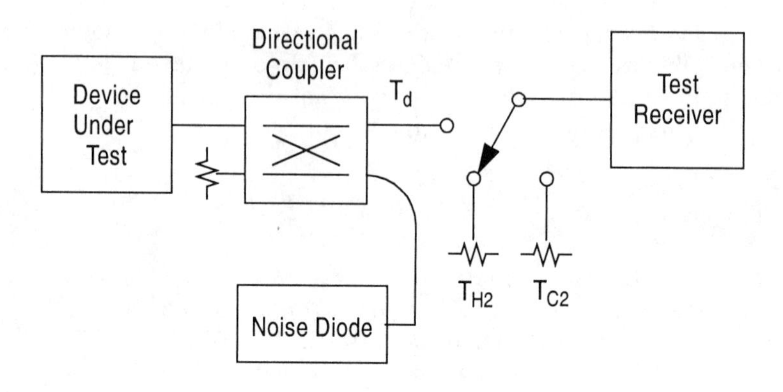

Figure 3.9 Test system for measuring reflection coefficient.

(3.53) and (3.54), which provide a general method for determining the noise temperature of a termination, broadly defined as either a resistive load or a port of a multiport component. When the noise source is fired, noise of temperature T_{di} is injected toward the DUT. The reflected noise, T_{dr}, is simply

$$T_{dr} = |\Gamma|^2 T_{di} \tag{3.55}$$

where Γ is the reflection coefficient of the port. If we know T_{di} and can measure T_{dr}, $|\Gamma|$ can be found directly.

T_{di} can be measured in a number of ways. The simplest is to replace the port of the DUT with a short circuit, fire the noise diode, and use (3.53) and (3.54) to determine the noise reflected from the short. When the DUT is connected, and the noise diode is fired, the noise temperature measured by the test receiver, T_{meas}, is

$$T_{meas} = T_L + T_{dr} \tag{3.56}$$

T_L, the output noise temperature of the DUT, can be found by the methods described in Section 3.3.2.

A disadvantage of this method is that the loss of the coupler is added to the measured gain of the DUT. This error can be minimized by the use of a low-loss coupler having approximately 20-dB coupling. If the noise source has a 25-dB or greater excess noise ratio (ENR; see Section 3.3.5), the value of T_{di} should be adequate for return-loss measurements to approximately 20 dB. The effect of the coupler's loss can then be removed from the gain and noise-temperature measurements by (3.8).

3.3.4 Sources of Error

Assessing the error from uncertainties in the noise-source temperatures or Y-factor measurements is straightforward. For example, to find the change in measured noise figure T_n from small errors in the hot-load temperature T_H, we need only differentiate (3.47) to obtain

$$\frac{\Delta T_n}{T_n} = \frac{T_H}{T_n}\frac{\partial T_n}{\partial T_H}\frac{\Delta T_H}{T_H} \tag{3.57}$$

We have expressed the error in T_n as a fractional quantity, and as a function of the fractional error in T_H. This case is usually most meaningful. Unfortunately, this exercise rarely results in expressions that offer much physical insight. It also does not identify other sources of error, which may be even more important. We describe a few of these below.

3.3.4.1 Test-System Stability

Throughout the description of the Y-factor measurement technique, we assumed that the gain of the test system does not change between the measurements. Any change in gain between the hot and cold measurements is reflected directly in an error in *Y*. Modern noise-test systems switch rapidly between hot and cold sources, minimizing such errors. When long integration times are needed to reduce the fluctuations in the measured noise level, it is better to switch fairly rapidly between the source and load, accumulating relatively short samples, than to dwell on one or the other for a long integration time.

3.3.4.2 Output-Level Fluctuations

The test system measures a mean-square voltage or current at its output, which is proportional to power. In the time domain, the mean-square current, $\overline{i^2(t)}$, is

$$\overline{i^2(t)} = \lim_{T \to \infty} \frac{1}{T} \int_0^T i^2(t) dt \tag{3.58}$$

To achieve an unambiguous measurement, it is necessary to integrate the noise over a theoretically infinite period of time. This is uncomfortably long.

The study of radiometers, of which our test system is an example, provides an expression for the noise uncertainty, ΔT, as a function of system temperature, predetection bandwidth, and postdetection integration time. That expression is

$$\Delta T = \frac{T_{sys}}{\sqrt{\Delta f \tau}} \tag{3.59}$$

where Δf is the RF predetection bandwidth and τ is the integration time, roughly the inverse of the postdetection bandwidth. T_{sys} is the system temperature, that is, the combined noise temperature of the sources, test receiver, and device under test. The RF bandwidth is often limited by the need to obtain adequate frequency resolution of the noise spectrum, so the integration time must be selected to make certain that noise uncertainties are adequately minimized.

3.3.4.3 Impedance Errors

Impedance errors are a significant source of error in noise measurements. The Y-factor measurement depends on accurate noise-source impedances, and especially on hot and cold standards having identical impedances. We have noted that the noise temperature of a two-port is fundamentally a function of source impedance; if the source impedances of the noise standards differ, an obvious uncertainty is introduced. Furthermore, from (3.20), the gain changes when the DUT is switched between loads, introducing an error in the Y factor.

3.3.4.4 Reflected Noise

Noise from the input of the test receiver is emitted toward the output of the DUT. That noise can be reflected from an imperfect output match into the input of the test receiver, changing its noise temperature. Reflected noise is rarely a significant problem in the measurement of amplifiers having more than a few decibels of gain, but in lossy components, such as mixers, it may be a much greater problem. Use of an isolator at the input of the test receiver minimizes the reflected-noise problem.

3.3.4.5 Spurious Signals

Spurious signals within the bandwidth of the measurement have much the same effect as additional noise in the DUT. Such signals can be generated within the test receiver or received by the DUT itself, often on power-supply leads, input leakage, or other such mechanisms. For this reason, measurement of very low-noise components or systems are often performed in a shielded enclosure.

The author vividly remembers one incident in which an anomalously high noise temperature in a very low-noise receiver was traced to stray pickup from the fluorescent ceiling lights in the laboratory.

3.3.4.6 Uncorrected Losses

Additional losses introduced in the test system from small coaxial or waveguide interconnections, or similar apparently negligible items, often have a surprisingly large effect on low-noise systems. The effect of such elements can be compensated either by modifying the noise-source temperature or using Friis' formula to remove their effect.

3.3.5 Noise Sources and Calibration

Two types of components are most frequently used as noise sources: cooled or heated resistors and electronic noise sources such as avalanche diodes. Other methods are possible. In the past, for example, vacuum diodes and gas-discharge tubes have been used as noise sources, but these are largely obsolete. Radio astronomers sometimes use well-characterized astronomical objects as noise standards for antenna and receiver measurements.

3.3.5.1 Excess Noise Ratio

Noise sources, especially noise diodes, are frequently specified by their *excess noise ratios* in decibels, or *ENR*. The ENR is defined as the ratio of the increase in noise over the standard temperature of $T_0 = 290\text{K}$, to the standard temperature itself. Specifically,

$$\mathrm{ENR} = 10 \log\left(\frac{T_n - T_0}{T_0}\right) \tag{3.60}$$

where T_n is the noise temperature of the source.

ENR is only marginally more useful than the noise temperature; for example, the ENR of a noise source followed by an attenuator, at 290K, is simply the ENR minus the attenuator loss in decibels. It exists in the engineering lexicon largely because of the preference for expressing virtually everything in logarithmic quantities.

3.3.5.2 Cooled or Heated Resistors

We have made the case that the available power from a matched termination is a function solely of its physical temperature. Therefore, heated or cooled resistors make ideal noise sources. In fact, they can be used as primary noise standards.

Section 3.3.1 shows that noise-figure measurements require at least two noise standards. A room-temperature resistor is an obvious choice for one such standard. A second resistor, cooled to a very low temperature, is another. A common coolant is liquid nitrogen, as it is inexpensive, easy to handle, provides a reasonably large $\Delta T = T_H - T_C$, and its boiling point (and therefore the resistor temperature) is well known, 77K at sea-level atmospheric pressure. The greatest problem in using liquid nitrogen is that many types of microwave terminations can be damaged by the thermal shock of sudden cooling to cryogenic temperatures.

Alternatively, a resistor can be heated. In this case it is necessary to make certain that the resistor is heated uniformly and its temperature is measured accurately. Heating allows for somewhat larger temperature difference ΔT than cooling, and therefore may provide improved accuracy for measurements of high noise temperatures. Its temperature also can be set to the desired value, a characteristic not practically possible with liquid-nitrogen-cooled terminations.

Determining the noise temperature of a cooled or heated coaxial termination is usually not too difficult. The accuracy of that temperature depends more on the characterization of the coaxial line connecting the load to the device under test than on the load itself. From (3.17), both the loss and the physical temperature must be known; however, the temperature is unlikely to be uniform along the line, and the loss depends on the temperature. From (3.17), the lower the overall loss in the line, the less that uncertainties in that loss and temperature affect the noise temperature of the load. Therefore, minimizing line loss also minimizes the uncertainties in the necessary corrections.

Liquid-nitrogen-cooled waveguide terminations require additional considerations. To prevent water condensation in the waveguide, a hermetically sealed window must be used at the output, and the load itself should be filled with a gas, such as helium, whose boiling point is below that of the liquid nitrogen. Radiative heat transfer from the external waveguide to the cooled element, whose heat-sinking to the liquid nitrogen is usually imperfect, can raise the termination's temperature. For this reason, it is sometimes wise to put a bend in the waveguide between the termination and the output port, so thermal radiation is directed toward a cold waveguide wall instead of the termination itself.

Change in impedance with temperature is a perennial problem in the design of terminations that are heated or cooled to extreme temperatures. Frequently, such terminations are damaged by heating or cooling. The damage may occur only after a number of temperature cycles and often can be found only from VSWR measurements.

3.3.5.3 Diode Noise Sources

Avalanche diodes are frequently used as sources of high-level noise, on the order of 5 to 25 dB ENR. Such sources generate shot noise from a dc excitation. From (2.53) and Figure 2.1, we can obtain an expression for the available noise power from an avalanche diode:

$$T_n = \frac{qIR}{2K} \tag{3.61}$$

which shows that 100 mA of dc current forced backwards through a long-suffering avalanche diode, in a 50Ω system, generates a noise temperature of ~29,000K.

Figure 3.10 shows the circuit of such a source. A well-regulated dc supply is necessary, as the noise temperature is proportional to the dc current. Of course, the diode capacitance must be low to facilitate a match, and the diode heat sinking must be adequate to prevent overheating. To assure good VSWR, an output attenuator is often used. Even with these precautions, the noise temperature is not sufficiently predictable to allow diode noise sources to be used as standards. Invariably, they are calibrated by comparison to thermal sources (Section 3.3.5.4).

The most common values of ENR for diode noise sources are 5.2 dB and 15.2 dB. A value around 25 dB is also encountered occasionally. These

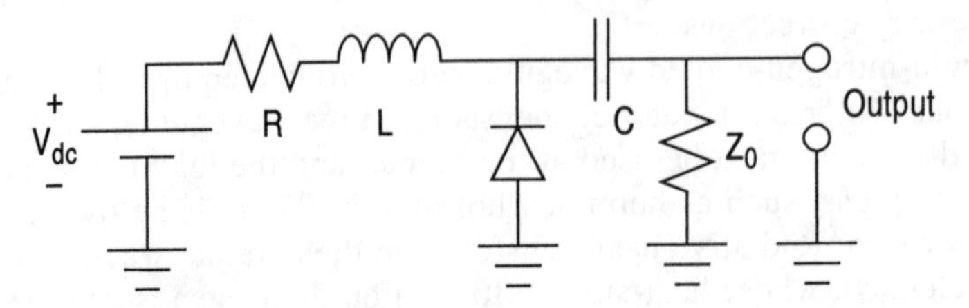

Figure 3.10 Circuit of an avalanche-diode noise source. The diode is operated in avalanche breakdown; the current is limited by resistor R. L and C provide RF and dc decoupling.

seemingly odd values were chosen long ago for compatibility with gas-discharge noise sources, whose ENRs were invariably close to 15.2 dB. Frequently a gas-discharge tube was padded with a 10 dB attenuator to provide a lower ENR value, 5.2 dB. Noise diodes also have output attenuators to optimize the VSWR and to minimize changes in VSWR when the diode is switched on or off.

3.3.5.4 Noise-Source Calibration

Diode noise sources can be calibrated by the noise measurement method described in Section 3.3.2. The test setup is shown in Figure 3.11. It is similar to the test receiver in Figure 3.8, with the noise diode replacing the output port of the DUT. Quantities are measured as in (3.53), with T_d replacing T_{o1}, and (3.54) is used to find T_d.

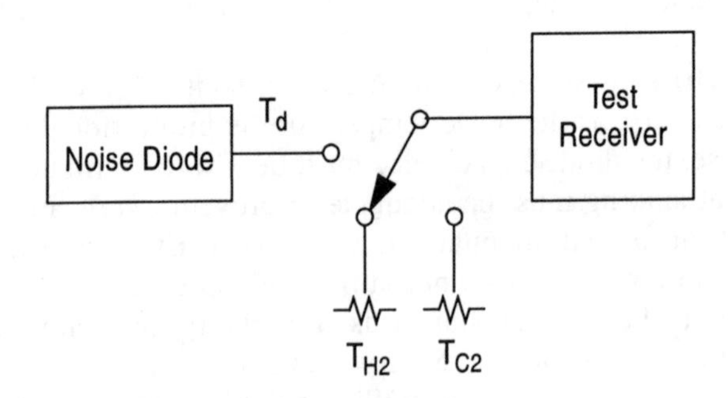

Figure 3.11 Noise diode calibration by comparison to known hot and cold loads.

References

[3.1] H. A. Haus, "Description of the Noise Performance of Amplifiers and Receiving Systems," *Proc. IEEE*, Vol. 51, p. 436, 1963.

[3.2] "IRE Standards on Electron Tubes: Definition of Terms, 1957," *Proc. IRE*, Vol. 45, p. 1000, 1957.

[3.3] "IRE Standards on Methods of Measuring Noise in Linear Two-Ports, 1959," *Proc. IRE*, Vol. 48, p. 60, 1960.

[3.4] H. A. Haus et al., "Representation of Noise in Linear Two-Ports," *Proc. IRE*, Vol. 48, p. 69, 1960.

[3.5] G. Gonzales, *Microwave Transistor Amplifiers*, Englewood Cliffs, NJ: Prentice-Hall, 1984.

[3.6] S. A. Maas, *Microwave Mixers*, Norwood, MA: Artech House, 1993.

[3.7] A. R. Kerr, "Shot Noise in Resistive-Diode Mixers and the Attenuator Noise Model," *IEEE Trans. Microwave Theory Tech.*, Vol. MTT-27, p. 135, 1979.

[3.8] C. Dragone, "Analysis of Thermal and Shot Noise in Pumped Resistive Diodes," *Bell Syst. Tech. J.*, Vol. 47, p. 1883, 1968.

Chapter 4

Noise Models of Solid-State Devices

In modeling noise in solid-state devices, we must address a number of matters. First, of course, is the nature of the sources and their location in the equivalent circuit or black-box model of a solid-state device. This is closely allied with the nature of the device's circuit model. The second is the scaling of the sources and their correlations with frequency and, in many cases, control voltages or currents. Finally, we usually must deal separately with low-frequency noise ($1/f$ and burst noise) and broadband, high-frequency noise, which have distinctly different physical sources and characteristics.

Perhaps because of the difficulty of these questions, few modern models in widespread use address them completely. At this writing, the RF and microwave industry still depends heavily on models introduced in the SPICE circuit simulator, which dates from the mid-1970s. In this chapter, we devote considerable effort to their description. Some of these models, such as the Schottky diode noise model, are surprisingly good; others less so.

It is important to recognize that SPICE does not include nonlinear noise analysis. Thus, the noise models developed for SPICE were originally intended for linear noise analysis at a bias point determined by SPICE's dc analysis. Their use for nonlinear noise analysis depends on an assumption that such static models are valid for dynamic applications as well.

4.1 RESISTORS AND PASSIVE, LOSSY ELEMENTS

The simplest noisy element to model is a resistor. We saw, in Section 2.3.1, that a resistor generated white thermal noise and that the available noise spectral power from a resistor was simply KT, where K is Boltzmann's constant and T is its absolute temperature in Kelvins. Because of this property, a noisy resistor can be modeled as a noiseless resistor having a series voltage noise source or a shunt current noise source. The mean-square value of the voltage, $\overline{|v_n|^2}$, is

$$\overline{|v_n|^2} = 4KTR\Delta f \tag{4.1}$$

For this and other kinds of noise sources, we sometimes express the noise as a spectral density, $S_v(f)$,

$$S_v(f) = 4KTR \tag{4.2}$$

The units of $S_v(f)$ are volts squared per hertz. Noise voltage spectra are also expressed occasionally in volts per root hertz.

When the resistor's noise is modeled as a shunt current source, the mean-square current, $\overline{|i_n|^2}$, is

$$\overline{|i_n|^2} = \frac{4KT\Delta f}{R} \tag{4.3}$$

and its spectral density follows similarly:

$$S_i(f) = \frac{4KT}{R} \tag{4.4}$$

Some kinds of resistors exhibit low-frequency $1/f$ noise. Such noise, when it occurs, is usually small compared to the low-frequency noise in solid-state devices, so it is rarely a concern.

All passive lossy elements are noisy. Such elements can be described by a correlation matrix (Section 2.2.4). The current correlation matrix, C_i, can be found from the admittance matrix, Y, of the component. It is [4.1]

$$C_i = 4KT\mathrm{Re}\{Y\} \tag{4.5}$$

Note that (4.5) requires that the element be passive and lossy, but it need not be reciprocal.

4.2 SCHOTTKY-BARRIER DIODES

The dominant high-frequency noise sources in Schottky-barrier diodes are shot noise in the junction and thermal noise in the series resistance. Schottky diodes also exhibit substantial levels of low-frequency noise, which consist of a combination of flicker ($1/f$) and burst noise. These are well modeled by the expressions in Sections 2.3.2 to 2.3.4. It has been shown that making the exponent in the denominator of the burst-noise expression a model parameter (2.55) can provide increased accuracy [4.2].

Figure 4.1 shows the equivalent-circuit and noise model of the diode. The current source $I_j(V_j)$ represents the exponential junction current, and the nonlinear capacitance, the junction charge. These are well described by the textbook equations,

$$I_j(V_j) = I_0 \exp\left(\frac{qV_j}{\eta KT}\right) \tag{4.6}$$

where q is electron charge, K is Boltzmann's constant, T is temperature in Kelvins, I_0 is a current parameter, sometimes called the *reverse saturation current*, and η is an empirical constant, usually in the range 1.15 to 1.25, called the *ideality factor*. The capacitance is described by its charge/voltage (Q / V) characteristic,

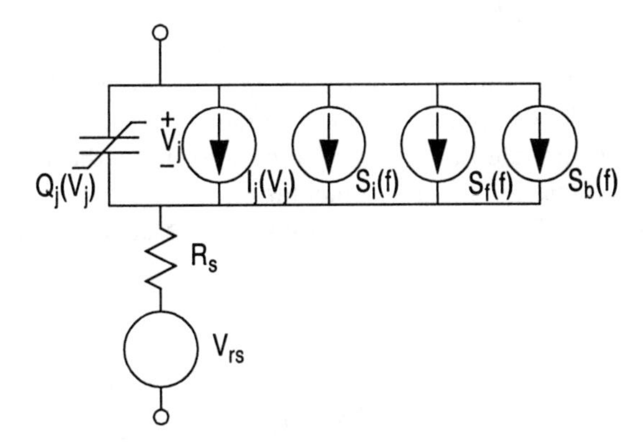

Figure 4.1 Diode equivalent circuit showing the noise sources for the junction and series resistance. The noise of the series resistance is expressed as a series voltage noise source, but it could also be shown as a shunt current noise source.

$$Q_j(V_j) = \frac{-C_{j0}\phi}{1-\gamma}\left(1 - \frac{V_j}{\phi}\right)^{1-\gamma} \tag{4.7}$$

where C_{j0} is the zero-voltage junction capacitance, ϕ is the junction potential, and γ is a constant, ideally 0.5 for uniformly doped diodes. This expression is valid when $V_j < \phi$ and at reverse voltages that do not fully deplete the diode's epitaxial layer. The diode also has a weakly nonlinear series resistance, which is usually treated as a linear resistance.

The expressions (4.6) and (4.7) have a number of numerical problems, some of which are subtle and others obvious. Equation (4.6) often requires greater numerical range than computers can provide, both at large forward and large reverse voltages, so it is necessary to create a weakly nonlinear extension at extreme junction voltages. Equation (4.7) has obvious problems when $V_j \geq \phi$. SPICE uses a quadratic extension of the Q/V characteristic above $F_c\,\phi$, where F_c is a model parameter, to avoid this difficulty. See [4.3, 4.4] for further details of the diode's electrical model.

The shot-noise and low-frequency noise sources are in parallel with the junction current source and are modulated by the junction current. The thermal noise source in series with the resistor is not modulated.

Low-noise diode mixers are still the dominant devices for submillimeter-wave receivers. In such applications they are often cooled to cryogenic temperatures, and in these extreme cases, other noise mechanisms become significant. See [4.5] for further discussion of this subject.

4.2.1 SPICE Noise Model

The original SPICE implementation of the diode-noise model included only shot noise, flicker noise, and, of course, thermal noise. Later implementations have included burst noise and have made the flicker-noise expression more versatile. The implementation in Microwave Office [4.6] is typical. It uses the following expressions for the spectral densities of the noise processes:

Shot noise:

$$S_i(f) = 2qI_j \tag{4.8}$$

Flicker noise:

$$S_f(f) = KF\frac{I_j^{AF}}{f^{FFE}} \tag{4.9}$$

Burst noise:

$$S_b(f) = KB\frac{I_j^{AB}}{1 + \left(\frac{f}{FB}\right)^2} \tag{4.10}$$

In the above expressions, *KF*, *KB*, *AF*, *FFE*, and *FB* are the SPICE parameter names, empirical constants in the equations. We show them as multi-character, upper-case terms, as these are the parameter names used in SPICE. The thermal noise of the series resistance, R_s, is modeled in the conventional manner (4.1).

4.3 JFET AND MOSFET NOISE MODELING

Because of the extraordinarily large numbers of published MOSFET models, it would seem difficult to develop generalizations about noise modeling of such devices. However, much of the modeling derives from early work on FET devices by van der Ziel, [4.7, 4.8], and this theory has influenced the development of FET noise models for many years.

4.3.1 Noise in Silicon FETs

In discussing noise in transistors, it is common to distinguish between *intrinsic* and *extrinsic* noise sources. Intrinsic sources are those associated with the basic device function, such as high-field diffusion noise in a FET's channel. Extrinsic sources are those associated with the transistor's external parasitic elements, such as thermal noise in the source, gate, and drain resistances or in a resistive substrate. When they amount to simple thermal noise, the characterization of extrinsic thermal noise sources is straightforward, so we shall be less concerned with those.

Modeling of silicon JFETs and MOSFETs has traditionally proceeded from the assumption of a fully resistive channel; that is, carriers do not reach saturated drift velocity, and therefore the incremental conductance of any longitudinal segment of the channel is nonzero. This is an underlying assumption of Shockley's original theory of FETs. Under this assumption, the device's noise is dominated by thermal noise in the channel. Later research showed that, even in silicon devices, carriers normally reach saturated drift velocity at the drain end of the channel, so this assumption is at best incomplete. It is especially questionable when applied to modern silicon MOSFETs and is completely unwarranted in III-V devices. It does, however, allow for good physical insight into FETs' thermal noise characteristics. Furthermore, other phenomena, such as high-field diffusion noise, can be included through extensions of the theory or as additional noise sources.

The current in the FET's channel can be expressed as

$$I_d = g(V_0)\frac{dV_0}{dx} \tag{4.11}$$

where $g(x) = g(V_0)$ is the conductance of an increment of the channel for unit length at point x (measured from the source contact) in Ω^{-1}cm, where the potential is V_0. Then

$$I_d dx = g(V_0)dV_0 \tag{4.12}$$

I_d, of course, must be constant along the channel, whose length is L. Then we can integrate both sides of (4.12) to obtain

$$I_d = \frac{1}{L}\int_0^{V_d} g(V_0)\,dV_0 \tag{4.13}$$

Thermal noise in the channel takes the form of a distributed, time-varying perturbation of V_0. We can then say $V = V_0 + \Delta V(x, t)$ and that the current is similarly $I_d + \Delta I_d(t)$. Then,

$$I_d + \Delta I_d(t) = g(V)\frac{dV}{dx} + i_d(x, t) \tag{4.14}$$

where $i_d(x, t)$ is the distributed noise current along the channel. Expanding and retaining only the first-order terms gives

$$\Delta I_d(t) = \frac{d}{dx}[g(V_0)\Delta V(x, t)] + i_d(x, t) \tag{4.15}$$

By treating the channel noise components as uncorrelated, and noting that the noise is thermal, one can show [4.8] that the noise spectral density of the drain current, $S_i(f)$, is

$$S_i(f) = \frac{4KT}{L^2 I_d}\int_0^{V_d} g^2(V_0)\, dV_0 \tag{4.16}$$

which is a white-noise process. Finally, if hot-electron effects are present, as is often the case in short-channel MOSFETs, the electron temperature becomes $T_e(x) = T_e(V_0)$ and (4.16) becomes

$$S_i(f) = \frac{4KT}{L^2 I_d}\int_0^{V_d} \frac{T_e(V_0)}{T} g^2(V_0) dV_0 \tag{4.17}$$

Equation (4.16) can be used in conjunction with the I/V model of any silicon FET to determine the channel's thermal noise. For example, a simple JFET model has

$$g(V_0) = g_0\left[1 - \left(\frac{V_{gs} + \phi + V_0}{V_p + \phi}\right)^{1/2}\right] \tag{4.18}$$

where ϕ is the built-in voltage of the gate-to-channel junction, g_0 is the zero-bias value of $g(V_0)$, V_p is the pinch-off voltage, and V_{gs} is the applied gate-to-source voltage. Substituting this expression into (4.16) results in the expression,

$$S_i(f) = \frac{2KTg_0}{L}\frac{(1 + 3z^{1/2})}{(1 + 2z^{1/2})} \tag{4.19}$$

where

$$z = \frac{V_{gs} + \phi}{V_p + \phi} \tag{4.20}$$

4.3.2 SPICE MOSFET Noise Model

SPICE, as originally developed by the University of California at Berkeley, contains three MOSFET models, designated as the level 1, 2, and 3 models. (Additional models have been added to later SPICE variants; one such variant, HSPICE, has over 50 MOSFET models.) Level 1 is a simple textbook model; level 2 is a largely unsuccessful attempt to overcome the limitations of level 1, and level 3 is a somewhat more successful attempt. The deficiencies of these models have been well documented; lack of charge conservation in the capacitance models and discontinuities in the I/V expressions are the primary problems. Implementations in other circuit simulators often include corrections for these problems, resulting in inconsistencies between supposedly identical models in different simulators. The SPICE noise model described in this section applies to all three levels.

Figure 4.2 shows the basic MOSFET equivalent circuit, including its I/V and noise sources. The low- and high-frequency noise sources are in parallel with the channel. Thermal noise sources associated with the resistive parasitics are included as well.

4.3.2.1 Channel Thermal Noise

The theory in Section 4.3.1 can be applied to MOSFETs in a straightforward manner. Since we assume the channel to be resistive over virtually its entire length, the noise is predominantly thermal. The channel noise is modeled in the circuit as a noise source in parallel with the channel's controlled current source. MOSFETs include additional thermal noise sources associated with their extrinsic resistive parasitics.

Simple MOSFET theory [4.3, 4.9] provides the induced channel charge per area, Q_{ch}:

$$Q_{ch}(V_0) = C_{ox}(V_{gs} - V_t - V_0) \tag{4.21}$$

where V_t is the threshold voltage and C_{ox} is the oxide capacitance, the capacitance per unit area of the capacitor formed by the gate and channel. Then, when $V_{gs} > V_t$,

$$g(V_0) = \mu W C_{ox}(V_{gs} - V_t - V_0) \tag{4.22}$$

where μ is mobility and W is the channel width. Substituting into (4.16) gives

$$S_i(f) = \frac{8KTg_0}{3L} \tag{4.23}$$

which is independent of V_{gs}. From (4.22) and (4.16) one can also derive an expression for g_0:

$$g_0 = \mu W C_{ox}(V_{gs} - V_t) = g_m L \tag{4.24}$$

Finally, substituting (4.24) into (4.23) gives the well-known SPICE expression for channel noise in MOSFETs:

$$S_i(f) = \frac{8KTg_m}{3} \tag{4.25}$$

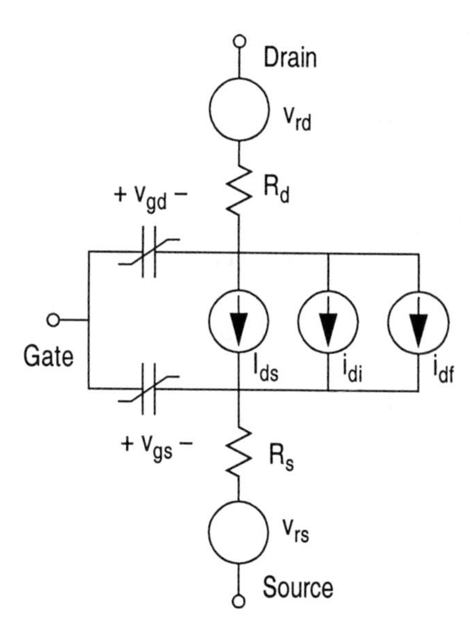

Figure 4.2 MOSFET equivalent circuit including noise sources. i_{di} and i_{df} are the channel high- and low-frequency noise sources, respectively. v_{rd} and v_{rs} are the thermal noise sources associated with the drain and source resistances. I_{ds} is the channel current source.

Sometimes the noise of the drain-to-source conductance g_{ds} and back-gating transconductance g_{mb} are included to form

$$S_i(f) = \frac{8KT(g_m + g_{mb} + g_{ds})}{3} \tag{4.26}$$

This expression is strictly valid only in current-saturated operation. The latter term is necessary, as (4.25) predicts zero noise when the drain-to-source voltage is zero and $g_m = 0$. Unfortunately, in linear operation, $S_i(f) = 4KTg_{ds}$, so (4.26) is still inconsistent between operating regions. Problems like this have stimulated further development of MOSFET noise models.

Many modern MOSFET models, however, use the same approach, substituting only more complex expressions for $g(V_0)$ [4.10]. Some use (4.25) as is, substituting only a more complex expression for g_m.

4.3.2.2 Low-Frequency Noise

The SPICE low-frequency noise model includes only flicker noise. It uses the same expression as for the diode:

$$S_f(f) = KF\frac{I_d^{AF}}{f^{FFE}} \tag{4.27}$$

where I_d is given in the references as the drain current; more precisely, it is the resistive component of the channel current. Some circuit simulators use the expression,

$$S_f(f) = \frac{KF}{C_{ox}L_{eff}W_{eff}}\frac{I_d^{AF}}{f^{FFE}} \tag{4.28}$$

while others offer

$$S_f(f) = \frac{KF}{C_{ox}L_{eff}W_{eff}}\frac{g_m^2}{f^{FFE}} \tag{4.29}$$

where L_{eff} and W_{eff} are the effective length and width of the gate, respectively. The latter expressions allow for scaling with device size, and offer slightly different dependence on channel current. Like the high-frequency noise source, the flicker noise source is in parallel with the channel.

4.3.2.3 Other Noise Sources

The SPICE noise model includes the noise in the extrinsic resistive parasitic drain and source resistances. These are modeled in the usual fashion, given by (4.1) to (4.4). MOSFETs exhibit other types of thermal and non-thermal noise, which are not included in the simple SPICE noise model, although noise models of advanced devices may include them. These are the following:

1. Thermal noise in other resistive parts of the device, including the gate and substrate;
2. Noise due to impact-ionization current in the channel at gate voltages well below threshold;
3. Shot noise associated with gate leakage current;
4. Shot noise from subthreshold conduction;
5. Shot noise in the substrate-to-drain and substrate-to-gate diode junctions;
6. Induced gate noise (see Sections 4.3.4 and 4.4.1.1);
7. High-field diffusion noise;
8. Thermal noise in the gate resistance. (Many FETs, especially ones designed for RF and microwave operation, have significant gate resistance, which generates significant thermal noise.)

Some of these sources may not be significant, as they occur only in certain kinds of devices or occur only under extreme operating conditions. We examine some of them further in sections describing other devices and models.

4.3.3 BSIM3 Noise Model

BSIM is a MOSFET model developed by the University of California at Berkeley. It is intended to be a standard model for use in research and industry [4.11, 4.12]. At this writing, the third-generation model, BSIM3, is in widest use. Its current (and probably last) version is 3.2.2. The next-generation model is predictably named BSIM4. BSIM5 is also in the works.

The BSIM3 model provides separate sets of equations for the FET under weak inversion, strong inversion, and accumulation operation. Similarly, the noise model uses separate equations in these modes of operation. Ideally, the possibility of discontinuities between regions is obviated by empirical smoothing and limiting functions.

BSIM3 includes two user-selectable noise models: the SPICE noise model described in Section 4.3.2 [specifically (4.25)] and an extended one. The latter employs the theory in Section 4.3.1, but the model is based on the more complex expression for incremental channel conductance $g(V_0)$ implicit in the BSIM3 model. It also includes a number of noise sources that are significant only in extreme conditions, such as shot noise in conducting substrate junctions.

The extended noise model uses the following expression for the channel noise spectral density:

$$S_i(f) = 4KT\frac{\mu_{eff}}{L_{eff}^2}|Q_{inv}| \tag{4.30}$$

where μ_{eff} is the effective mobility of channel electrons, L_{eff} is the channel's effective length, and Q_{inv} is the total inversion charge in the channel. BSIM3 calculates these quantities as part of the model, for each region of operation, and for each of its several charge-model options. This expression is equivalent to $S_i(f) = (8/3)KT(g_m + g_{mb})$ for long-channel devices in saturated operation, but smoothly varies to $S_i(f) = 4KTg_{ds}$ in the linear region. Thus, it avoids the inconsistency described at the end of Section 4.3.2.1.

The low-frequency noise model similarly offers two options. The first is similar to (4.28):

$$S_f(f) = \frac{KF}{C_{ox}L_{eff}^2}\frac{I_d^{AF}}{f^{EF}} \tag{4.31}$$

in which EF is a model parameter. Equation (4.28) includes the term $L_{eff}W_{eff}$ while (4.31) uses L_{eff}^2. Since these quantities are approximately constant, the difference can be absorbed into the model parameter KF.

The extended low-frequency noise model includes a quadratic dependence on drain current. The expression is fairly complicated, as it includes terms dependent on other quantities calculated in the model. Most important is its dependence on a function of channel charge density, which in turn is a function of gate-to-source voltage, V_{gs}. The expression has the form

$$S_f(f) = K_1 \frac{I_d}{f^{EF}} f_1(V_{gs}) + K_2 \frac{I_d^2}{f^{EF}} f_2(V_{gs}) \tag{4.32}$$

where K_1 and K_2 are constant functions of physical and device parameters, and f_1 and f_2 are functions of V_{gs}. Specific information regarding the implementation can be found in [4.11].

BSIM3 does not include hot-electron modeling or induced gate-noise current. Since the model does not include an extrinsic gate resistance, it does not account for gate-resistance noise. These, plus the fact that it is based on the same fundamental theory as much older models, are significant deficiencies.

4.3.4 BSIM4 Noise Model

BSIM4 includes two noise models: a slightly modified version of the BSIM3 model and a new model, which its developers call the *holistic thermal noise model*.[1] In the former model, (4.30) is modified to

$$S_i(f) = \frac{4KT\,N_T}{r_{ds} + L_{eff}^2/(\mu_{eff}|Q_{inv}|)} \tag{4.33}$$

where N_T is a constant model parameter (called *NTNOI* in the model) and r_{ds} is the drain-to-source resistance, whose effect was not included in (4.30). The new parameter, N_T, provides an increase in the noise level for hot-electron effects, but it is clearly not scalable.

The model includes a channel noise source and an induced gate-noise source. The channel noise source includes g_m and g_d terms; in normal, current-saturated operation, it can be expressed approximately as

$$S_i(f) \sim 4KT \frac{V_{ds}}{I_d} K_1 g_m^2 \tag{4.34}$$

where K_1 is calculated from other quantities in the model. This seems to follow the earlier approach of multiplying the temperature in the thermal noise equations to account for high-field diffusion noise. The gate-noise current source is in parallel with the extrinsic source resistance, R_s. This is not a simple thermal resistive noise source; its value is calculated from in-

1. The author fervently hopes that this name is just a failed attempt at humor.

ternal quantities in the model. These noise sources are uncorrelated and frequency dependent. The model is similar in many ways to the Pospieszalski model, described in Section 4.4.3.

The details of this model are too complex to be addressed here. See [4.12] for further information.

4.4 MESFET AND HEMT NOISE MODELS

The operation of MESFETs and HEMTs is superficially similar to that of JFETs and MOSFETs but different in more fundamental characteristics. These differences in operation account for significant differences in noise properties.

In contrast to MOSFETs, the electrons in MESFETs and HEMTs move at saturated drift velocity over most of the channel. Nonthermal noise is generated in this region, where $g(V_0) = 0$. The dominant mechanism is high-field diffusion noise. The channel still includes a resistive portion, modeled by the so-called *intrinsic resistance* in most FET models. This is an important source of noise as well.

One important difference between MOSFETs and MESFETs or HEMTs is the significance of induced gate noise. Although it is clear that gate noise exists, the physical interpretation of that noise is less clear. Pucel et al. attribute it to the coupling of nonthermal channel noise to the gate, while Pospieszalski identifies a fundamentally thermal source. Most subsequent models have followed Pospieszalski's lead.

4.4.1 Noise Sources in Microwave FETs

4.4.1.1 Induced Gate Noise

A FET or bipolar transistor operated as a linear amplifier is used as a two-port device, in which the gate/base terminal is the input, the drain/collector is the output, and the source/emitter is common to both ports. We shall see in Chapter 5 that a noisy linear two-port is equivalent to a noiseless two-port with two correlated noise sources: one in shunt with the input and the other in shunt with the output.[2] These sources are equal to the short-circuit noise current at each port. The equivalence is illustrated in Figure 4.3.

So far, our MOSFET noise models have consisted of only a single noise source, a channel noise source that creates a noise output current, but

2. Other configurations are possible. For now, however, we limit our consideration to input and output shunt current sources.

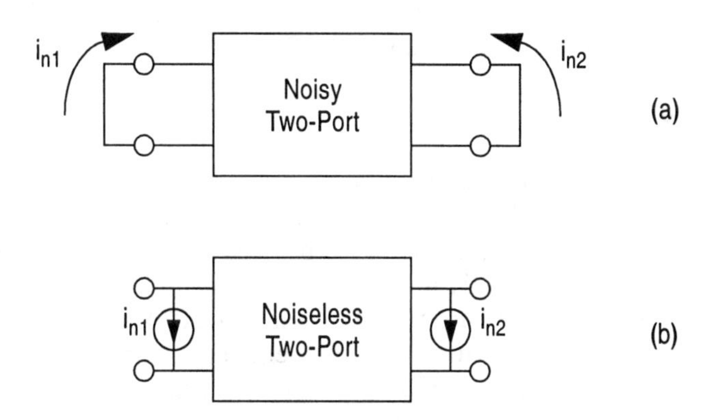

Figure 4.3 A noisy linear two-port can be modeled as a noiseless two-port having input and output noise sources. The short-circuit noise currents of the noisy two-port (a) are modeled in the noiseless two-port (b) by external noise sources. Because the port noise currents arise from the same internal sources, they are in general correlated.

virtually no input current. Without an input source, the model seems incomplete. Nevertheless, because of the high gate impedance of conventional MOSFETs in the frequency range where they are normally used, little of the channel noise can be coupled into the gate. Therefore, the short-circuit gate noise is often negligible, and the model, in the form of Figure 4.3, has no noise source at the input port.

One must ask, however, what happens at frequencies where the gate impedance is no longer high. This is often the case in advanced, short-channel MOSFETs and is invariably true of GaAs MESFETs and HEMTs. In these devices, noise from the channel can be coupled into the gate, through the gate-to-channel capacitance, resulting in a nonnegligible gate-noise current. This noise, called *induced gate noise*, results from the same noise sources as the drain noise, and thus is correlated with the drain noise. The nature of gate noise and its correlation with drain noise has been the subject of much research.

4.4.1.2 High-Field Diffusion Noise

Fairly early in the development of GaAs MESFET models it became clear that over a large portion of the FET's channel—in fact, over most of it—electrons move at saturated drift velocity. In this case, thermal noise from the channel is relatively small, and high-field noise dominates. This type of noise has been discussed in Section 2.3.4.2.

4.4.1.3 Low-Frequency Noise

Microwave FETs exhibit relatively high levels of low-frequency noise, $1/f$ noise generated in the channel. Phase noise in oscillators consists primarily of upconverted low-frequency noise, so devices having high levels of low-frequency noise are generally not preferred for use in low-noise oscillators. Bipolar and HBT devices are usually used instead. However, microwave FETs are capable of operation at higher frequencies than bipolars, so they are often the only option for millimeter-wave oscillators.

Conversely, unbiased FETs provide low $1/f$ noise when used as resistive mixers. This is especially valuable in receivers that convert directly to baseband. The $1/f$ noise level of such mixers is lower than comparable diode or bipolar mixers.

Low-frequency noise in microwave FETs is largely flicker noise, which is modeled in the conventional manner (Section 4.3.2.2).

4.4.2 Fukui Model

Early in the development of GaAs MESFETs, there was a clear need to express the minimum noise figure of a MESFET as a function of its macroscopic characteristics. One such method was developed by Fukui [4.13] and was known generally as the *Fukui equation*. This equation has largely been supplanted by later work, but because of its historical importance, we mention it here.

Fukui showed that the minimum noise figure of a MESFET could be given, to remarkably good accuracy, by

$$F_o = 1 + K_l L f \sqrt{g_m (R_g + R_s)} \tag{4.35}$$

where F_o is the optimum noise figure, L is the gate length, g_m is the transconductance, and R_g and R_s are the gate and source resistances, respectively. Most interesting is the empirical constant K_l, which must be determined for any FET fabrication process.

Fukui theory, while having some theoretical underpinnings, depends heavily on the Fukui constant, K_l, for its validity. Thus, it is largely an empirical expression useful for prediction and scaling. That is, knowing the value of K_l and some basic FET parameters, one can use (4.35) to scale simple noise-figure measurements to higher frequencies and different geometries.

4.4.3 Pospieszalski Model

Another linear noise model is that of Pospieszalski [4.14]. This model is designed to be scalable over frequency and temperature, but not to be a nonlinear noise model. However, by making its parameters functions of control voltages or currents in the model, it conceivably could be converted into a nonlinear model.

Pospieszalski showed that the intrinsic noise sources of a FET could be modeled by two parameters: the physical temperature, T_g, of the intrinsic resistance, R_i, and an elevated noise temperature, T_d, in the neighborhood of a few thousand Kelvins, attributed to the drain-to-source conductance, g_{ds}. T_g is, in all cases, close to the physical temperature of the device. We show this model in Figure 4.4. These noise temperatures are constant with frequency and the noise sources are uncorrelated. When drain current is appropriately scaled, T_d is not constant with physical temperature. Between 300K and 12.5K, for example, it varies by a factor of approximately 4.

If some reasonable conditions are met, simple expressions can be derived for the optimum source impedance and minimum noise figure. The condition for validity is

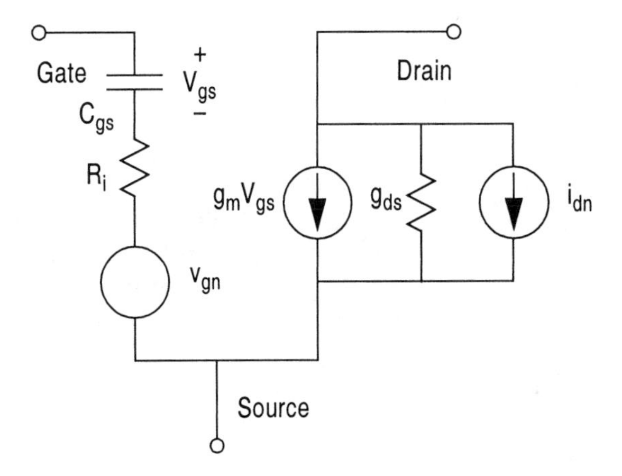

Figure 4.4 Equivalent circuit of the intrinsic FET used for the Pospieszalski noise model. $\overline{|v_{gn}|^2} = 4KT_g\Delta f R_i$ and $\overline{|i_{dn}|^2} = 4KT_d\Delta f g_{ds}$ are the noise sources. R_i and g_{ds} are noiseless resistances.

$$\frac{f}{f_t} \ll \sqrt{\frac{T_g}{T_d}} \tag{4.36}$$

then the minimum noise figure, $F_{\min}$, and optimum source impedance, Z_{opt}, are

$$F_{\min} \approx 1 + \frac{2}{T_0}\frac{f}{f_t}\sqrt{g_{ds}R_i T_g T_d} \tag{4.37}$$

$$Z_{\text{opt}} \approx \frac{f_t}{f}\sqrt{\frac{R_i T_g}{g_{ds}T_d}} - \frac{1}{j\omega C_{gs}} \tag{4.38}$$

The noise conductance, G_n, is given in all cases by

$$G_n = \left(\frac{f}{f_t}\right)^2 g_{ds}\frac{T_d}{T_0} \tag{4.39}$$

As is usual, $f_t = g_m / C_{gs}$. Note also a fundamental limitation on noise parameters:

$$\frac{T_{\min}}{T_0} \le 4G_n \text{Re}\{Z_{\text{opt}}\} \tag{4.40}$$

where $T_{\min}$ is the minimum noise temperature.

One of the problems in this model is the dependence of gate noise temperature on R_i, a parameter that is notoriously difficult to measure. In fact, it may be better to determine R_i by first assuming its noise temperature to be equal to its physical temperature, then extracting its value from noise measurements.

This model has been the subject of considerable research interest. As a result, it has been validated for a large number of devices, frequency ranges, and temperatures.

4.4.4 Pucel Model

A model developed by Pucel, Haus, and Statz, generally known as the *Pucel model*, was the first major comprehensive model of the GaAs

MESFET [4.15]. This model resolved a number of important questions about the operation and noise characteristics of such devices, and thus has had a strong influence on the subsequent development of FET noise theory. The model draws much of its inspiration from van der Ziel's earlier work involving thermal and nonthermal noise in FETs but includes much original theory as well. An important part of the model is its treatment of induced gate noise and the correlation coefficient between the gate and drain noise sources.

The Pucel model is fundamentally a MESFET model, as HEMTs did not exist at the time of its development. Still, since variations on the model are available in circuit simulators, it has been used regularly to model HEMTS as well. The model is quasiphysical; that is, it is based on a physical analysis of the device, but in practice the major parameters are determined empirically.

The Pucel model is fundamentally a small-signal noise model. However, because it includes parameters that are implicitly functions of internal control voltages and currents in the FET equivalent circuit, it can be used in large-signal circuit simulators for nonlinear noise analysis.

The small-signal equivalent circuit of the intrinsic device (not including gate-to-drain capacitance or resistive parasitics) is shown in Figure 4.5. Its noise sources are a drain noise source in parallel with the channel and a gate noise source in parallel with the gate-to-source junction. The latter

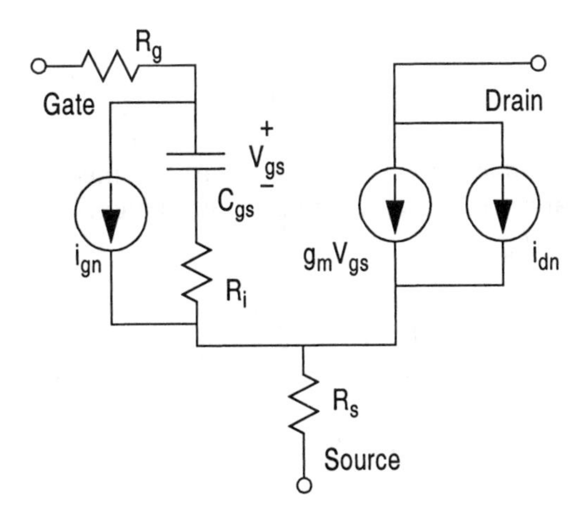

Figure 4.5 Equivalent circuit used for the Pucel model. The gate and source resistances, R_g and R_s, are explicitly included in the model.

source is nonwhite; this characteristic complicates its use in nonlinear noise analysis. Pucel et al. show that the correlation coefficient between these sources is imaginary.

The model uses three parameters, C, R, and P. C is the magnitude of the correlation coefficient, P is the normalized, dimensionless term representing the drain noise, and R is the analogous quantity for the gate noise. These are normalized in such a way that they should be at least approximately frequency independent. Reference [4.15] gives fairly complex expressions for these quantities, but in practice they are usually determined empirically. The quantities are defined as

$$P = \frac{\overline{|i_d|^2}}{4KT_0 g_m \Delta f} \tag{4.41}$$

$$R = \frac{\overline{|i_g|^2}}{4KT_0 \omega^2 \left(\dfrac{C_{gs}^2}{g_m}\right) \Delta f} \tag{4.42}$$

and

$$jC = \frac{\overline{i_g^* i_d}}{\sqrt{\overline{|i_d|^2}\,\overline{|i_g|^2}}} \tag{4.43}$$

(Note that, in accordance with our earlier warning, this is the complex conjugate of our standard definition of noise correlation.) The resulting minimum noise figure $F_{\min}$ is

$$\begin{aligned} F_{\min} = {} & 1 + 4\pi f \frac{C_{gs}}{g_m} \sqrt{K_g [K_r + g_m (R_g + R_s)]} \\ & + 8\left(\pi f \frac{C_{gs}}{g_m}\right)^2 K_g g_m (R_g + R_s + K_c R_i) + \ldots \end{aligned} \tag{4.44}$$

where

$$
\begin{aligned}
K_g &= P[(1 - C\sqrt{R/P})^2 + (1 - C^2)(R/P)] \\
K_r &= \frac{R(1 - C^2)}{(1 - C\sqrt{R/P})^2 + (1 - C^2)(R/P)} \\
K_c &= \frac{1 - C\sqrt{R/P}}{(1 - C\sqrt{R/P})^2 + (1 - C^2)(R/P)}
\end{aligned}
\tag{4.45}
$$

4.4.5 Chalmers Model

The dominant nonlinear circuit model for HEMTs was developed at Chalmers University in Göteborg, Sweden by I. Angelov et al. [4.16]. More recently, a noise model was included [4.17]. The equivalent circuit for the intrinsic FET with the model's noise sources is shown in Figure 4.6. The noise model follows the lead of Pospieszalski, consisting of uncorrelated drain-noise and gate-noise sources. Unlike Pospieszalski, however, the gate-noise source is in shunt with the gate-to-source capacitance. As with other models, the drain noise source is in shunt with the channel-current source. The model includes $1/f$ noise and shot noise in the gate, which can occur when the gate-to-channel junction is forward biased in large-signal operation. This is potentially an important noise source in oscillators and other large-signal circuits.

The drain and gate noise sources have the following spectral densities:

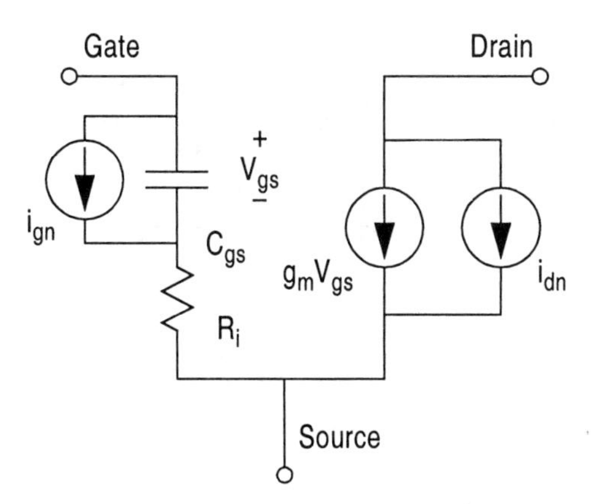

Figure 4.6 Equivalent circuit of the intrinsic device used in the Chalmers model.

$$I_{dt} = |I_{ds}| + |I_{gd}|$$

$$S_{id}(f) = 4KTt_{mn}\sqrt{I_{dt} + K_{nd2}I_{dt}^2}(1 + K_{lfd})W \cdot 10^{-2} \tag{4.46}$$

$$S_{ig}(f) = 4KT(I_g + I_{dt}/K_{ng1})(1 + K_{lfg})W \cdot 2 \cdot 10^{-2}$$

where I_{ds} and I_{gd} are the drain-to-source and gate-to-drain currents, respectively, and t_{mn}, K_{nd2}, and K_{ng1} are fitting parameters determined empirically. W is gate width in millimeters. The numerical constants and the use of millimeters for the gate width are not strictly necessary; they could be absorbed into other constants. K_{lfd} and K_{lfg} provide the low-frequency noise components. Since the frequency dependence of the sources is much the same, they can be treated by the same functions:

$$K_{lfg} \approx K_{lfd} = K_{lf}\left(\frac{1}{f^n} + \frac{B}{1 + \left(\frac{f}{f_{gr}}\right)^2}\right) \tag{4.47}$$

where f_{gr} and n are fitting parameters. The gate diode noise must also be included in cases where rectification in the gate-to-channel junction generates significant current:

$$S_{igr}(f) = 2qI_{gr} + K_f\frac{I_{gr}^{AF}}{f} \tag{4.48}$$

where I_{gr} is the rectified gate current in the gate-to-channel junction. This is essentially the diode noise model of (4.8) and (4.9).

It is interesting to note that this model gives a different short-circuit gate-noise component from the Pospieszalski model. In the Pospieszalski model, the short-circuit gate current, based on the intrinsic elements only, is

$$\overline{|i_g|^2} = \frac{\overline{|v_{gn}|^2}(C_{gs}\omega)^2}{(R_iC_{gs})^2 + 1} \tag{4.49}$$

while the Chalmers model gives

$$\overline{|i_g|^2} = \frac{\overline{|i_{gn}|^2}}{(R_i C_{gs})^2 + 1} \tag{4.50}$$

where i_g is the short-circuit gate-noise current and v_{gn}, i_{gn} are the noise sources of the model. In some cases, it may be necessary to use a nonwhite expression for either v_{gn} or i_{gn} to model the noise accurately over a broad bandwidth.

4.5 BIPOLAR AND HBT NOISE MODELS

By far, the dominant tool for modeling bipolar transistors has been the SPICE Gummel-Poon model (SGP). This model consists of the basic model published in 1970 by H. K. Gummel and H. C. Poon [4.18], enhanced by a number of extensions, one of which is the noise model. The dominance of the model can be attributed to its history; it was, after all, the first reasonably complete bipolar transistor model, and by now is available in literally every circuit simulator. Later models, while more accurate, are not yet in such wide use.

Homojunction and heterojunction devices are structurally similar, have similar noise sources, and thus are modeled similarly. High-frequency noise is dominated by shot noise, while flicker and burst noise dominate the low-frequency noise. Low-frequency noise is generally lower in bipolar devices than in FETs, making them preferable for applications in low-noise oscillators and baseband circuits.

4.5.1 BJT and HBT Noise Sources

The dominant noise sources in bipolar devices are shot noise in the junctions and thermal noise in the extrinsic parasitic resistances. Bipolars exhibit low-frequency noise similar to that of diodes. Like diodes, they exhibit burst noise. Although the original SPICE model does not include burst noise, most nonlinear circuit simulators include it as an extension.

Extrinsic noise sources have a strong effect on the noise temperatures of bipolars used in linear applications. The base and emitter resistances have an especially strong effect. Much of the literature concerning noise modeling of bipolars focuses on these resistances.

4.5.2 SPICE Gummel-Poon Noise Model

It is important to recognize that the SPICE noise models are linear, small-signal models, scalable according to bias. Implementing them in nonlinear analysis as pumped, time-varying noise sources is an extension of the models.

SPICE uses separate equivalent circuits for small-signal linear and large-signal nonlinear analyses. Since SPICE does not include nonlinear noise analysis, the noise equivalent circuit is part of the small-signal linear equivalent circuit only. For use in both linear and nonlinear noise analysis, it therefore must be generalized.

The small-signal noise model in SPICE consists of (1) a shot-noise source between the collector and emitter and (2) a combined shot and low-frequency noise source between the base and emitter. The collector noise spectrum is

$$S_{ic}(f) = 2qI_c \tag{4.51}$$

and the base noise is

$$S_{ib}(f) = 2qI_b + KF\frac{I_b^{AF}}{f} \tag{4.52}$$

where I_c and I_b are the resistive components of the collector and base currents, respectively, and the other parameters are the same as in the diode model. For small-signal linear analysis, I_c and I_b are dc bias currents; for large-signal analysis, they should be viewed as the resistive components of the junction currents.

Extending this model to the large-signal case is straightforward. Figure 4.7 shows the equivalent circuit of the bipolar device, including its noise sources. In that figure, the diodes represent the components of the base current. Two diodes are shown in each branch; one represents a variety of effects that we shall loosely call leakage current; the other, the base current generated by hole injection into the base. The collector current is the sum of currents in both junctions:

$$\begin{aligned} I_c &= I_{cc} - I_{ce} \\ &= \frac{IS}{Q_b}\left(\exp\left(\frac{qV_{be}}{NF \cdot KT}\right) - \exp\left(\frac{qV_{bc}}{NR \cdot KT}\right)\right) \end{aligned} \tag{4.53}$$

where *IS*, *NF*, and *NR* are the SPICE parameters for the junction reverse saturation current and ideality factors, respectively, and Q_b is the normalized minority base charge. Q_b, a function of both V_{be} and V_{bc}, accounts for Early effect and high-level injection. For the purposes of noise analysis, the current component I_{ce} should be viewed as a reduction in I_{cc}, not as a separate current component. Thus, the collector shot-noise current is a function of I_c only.

The base current is given by

$$\begin{aligned} I_{be} &= \frac{IS}{Q_b BF}\left(\exp\left(\frac{qV_{be}}{NF \cdot KT}\right) - 1\right) + ISE\left(\exp\left(\frac{qV_{be}}{NE \cdot KT}\right) - 1\right) \\ I_{bc} &= \frac{IS}{Q_b BR}\left(\exp\left(\frac{qV_{bc}}{NR \cdot KT}\right) - 1\right) + ISC\left(\exp\left(\frac{qV_{bc}}{NC \cdot KT}\right) - 1\right) \end{aligned} \tag{4.54}$$

where *BF* and *BR* are the forward and reverse current gains, SPICE BJT model parameters. Other variables in uppercase are also SPICE parameters.

In characterizing the noise sources for nonlinear analysis, we must accommodate inverse operation and cleanly separate the base and collector current components. This results in the same expression for collector noise, (4.51); the base current noise has two components, $S_{ibe}(f)$ and $S_{ibc}(f)$:

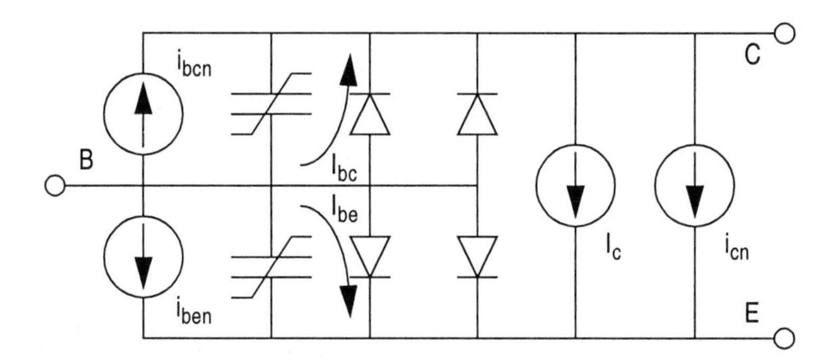

Figure 4.7 Intrinsic BJT with its noise sources. The base current components that generate noise I_{bc} and I_{be} are only the components in the diodes. i_{ben} and i_{bnc} are the respective base-noise components. i_{cn} is the collector noise source.

$$S_{ibe}(f) = 2qI_{be} + KF\frac{I_{be}^{AF}}{f} \tag{4.55}$$

$$S_{ibc}(f) = 2qI_{bc} + KF\frac{I_{bc}^{AF}}{f} \tag{4.56}$$

Because of its small value in normal operation, the base-to-collector component is often neglected. The current components are shown in Figure 4.7.

Van der Ziel [4.8] shows that these expressions are an approximation, as they do not include the reverse current components in the junction. In theory, these currents give the base and collector noise currents a weak correlation, as they are common to both. In practice, the noise components and their correlation are invariably negligible.

4.5.3 Noise Characterization in Advanced Bipolar Models

At this writing, the dominant advanced models of bipolar devices are VBIC [4.19], MEXTRAM [4.20], and HICUM [4.21]. These were originally developed for homojunction transistors. Although the models are used regularly for heterojunction devices, models specifically developed for HBTs have come from Chalmers University [4.22] and the University of California at San Diego (UCSD) [4.23].

The noise formulation described in Section 4.5.2 is the basis for all these advanced models. Most of the advanced models include additional current components, which have their own shot-noise and flicker-noise sources, and more complete treatment of the SGP noise sources. In the UCSD HBT model, for example, the low-frequency noise parameters *KF*, *AF*, and BFN [which is equivalent to *FFE* in (4.9)] are temperature dependent. The reader should consult the references for details of these implementations.

References

[4.1] R. Q. Twiss, "Nyquist's and Thevenin's Theorems Generalized for Nonreciprocal Linear Networks," *Journal of Applied Physics*, Vol. 26, p. 599, 1955.

[4.2] K. F. Sato et al., "Low-Frequency Noise in GaAs and InP Schottky Diodes," *IEEE MTT-S International Microwave Symposium Digest*, 1998.

[4.3] G. Massobrio and P. Antognetti, *Semiconductor Device Modeling with SPICE*, New York: McGraw-Hill, 1993.

[4.4] S. A. Maas, *Nonlinear Microwave and RF Circuits*, Norwood, MA: Artech House, 2003.

[4.5] G. Hegazi, A. Jelenski, and S. Yngvesson, "Limitations of Microwave and Millimeter-Wave Mixers Due to Excess Noise," *IEEE MTT-S International Microwave Symposium Digest*, p. 431, 1985.

[4.6] Applied Wave Research, 1960 E. Grand Avenue, El Segundo, CA 90245.

[4.7] A. van der Ziel, *Noise: Sources, Characterization, Measurement*, Englewood Cliffs, NJ: Prentice-Hall, 1970.

[4.8] A. van der Ziel, *Noise in Solid State Devices and Circuits*, New York: Wiley, 1986.

[4.9] D. Foty, *MOSFET Modeling with SPICE*, Upper Saddle River, NJ: Prentice-Hall, 1997.

[4.10] H. C. de Graaf and F. M. Klaasen, *Compact Transistor Modeling for Circuit Design*, Vienna: Springer-Verlag, 1990.

[4.11] Y. Cheng and C. Hu, *MOSFET Modeling and BSIM3 User's Guide*, Boston: Kluwer, 1999.

[4.12] W. Liu, *MOSFET Models for SPICE Simulation*, New York: Wiley, 2001.

[4.13] H. Fukui, "Optimal Noise Figure of GaAs MESFETs," *IEEE Trans. Electron Devices*, Vol. ED-26, p. 1032, 1979.

[4.14] M. W. Pospieszalski, "Modeling of Noise Parameters of MESFETs and MODFETs and Their Frequency and Temperature Dependence," *IEEE Trans. Microwave Theory Tech.*, Vol. 37, p. 1340, 1989.

[4.15] R. A. Pucel, H. A. Haus, and H. Statz, "Signal and Noise Properties of Gallium Arsenide Microwave Field Effect Transistors," *Advances in Electronics and Electron Physics*, Vol. 38, New York: Academic Press, 1975.

[4.16] I. Angelov, H. Zirath, and N. Rorsman, "A New Empirical Nonlinear Model for HEMT and MESFET Devices," *1992 Transactions on Microwave Theory and Techniques,* Vol. 40, p. 2258, 1992.

[4.17] I. Angelov, R. Kozhuharov, and H. Zirath, "A Simple Bias Dependent LF FET Noise Model for CAD," *IEEE MTT-S International Microwave Symposium Digest*, 2001.

[4.18] H. K. Gummel and H. C. Poon, "An Integral Charge-Control Model of Bipolar Transistors," *Bell Sys. Tech. J.*, Vol. 49, No. 3, p. 827, 1970.

[4.19] C. McAndrew et al., "VBIC95, The Vertical Bipolar Inter-Company Model," *IEEE J. Solid-State Circuits*, Vol. 31, p. 1476, 1996.

[4.20] H. C. de Graaf and F. M. Klaasen, *Compact Transistor Modelling for Circuit Design*, New York: Springer-Verlag, 1990.

[4.21] H.-M. Rein et al., "A Semi-Physical Bipolar Transistor Model for the Design of Very High-Frequency Analog ICs," *Proc. IEEE Bipolar and BiCMOS Circuits and Technology Meeting*, p. 217, 1992.

[4.22] I. Angelov, "A Simple HBT Large-Signal Model for CAD," *IEEE MTT-S International Microwave Symposium Digest*, 2002.

[4.23] UCSD Electrical Engineering Dept., High-Speed Devices Group, *HBT Modeling, rev. 9.001A*, http://hbt.ucsd.edu, March 2000.

Chapter 5

Noise Theory of Linear Circuits

Analysis of noise in linear circuits is probably the subject of greatest interest in low-noise design. In this chapter, and in contrast to the treatment in Chapter 3, we approach the subject from a circuit, rather than a system, viewpoint. Much of the basic work in this area was performed quite a long time ago, so the chapter draws heavily on the work listed in the references, especially [5.1–5.3]. The methods are designed mainly for efficiency in performing noise analysis by computer, as this is the primary application of the theory.

5.1 NOISE SOURCES

In Chapter 2 we saw that noise can be generated by a number of different mechanisms: thermal noise, shot noise, flicker noise, and so on. In this chapter, we are not concerned with the physical source of the noise, but instead we need only know that it exists. In all cases, we deal with a narrow spectral segment of high-frequency, white noise. Generally, that noise is modeled as a current or voltage source. Each source may be correlated with one or more other sources, which are not necessarily of the same type.

The white-noise assumption is not particularly limiting. The bandwidth of interest can always be made narrow enough to make the noise spectral density constant. In many cases, that bandwidth may be as narrow as 1 Hz, but in general it will not be stated explicitly; noise quantities in this chapter can generally be viewed as spectral densities, or, if the reader demands something more concrete, as power in a 1-Hz bandwidth. We can treat non-white noise simply by analysis over a range of frequencies, with the proper noise level at each point. In essence, we sample the noise spectrum over that range of frequencies.

Under these assumptions, the noise of each source can be viewed as a quasisinusoid, a sine wave varying slowly in magnitude and phase. The mean-square power of a current noise source is simply that of the sinusoid, $\overline{i^2(t)}$:

$$\overline{|i|^2(t)} = \overline{ii^*}\Delta f = \overline{|i|^2}\Delta f = \lim_{T \to \infty} \frac{1}{T}\int_0^T i^2(t)dt \tag{5.1}$$

where i (without the time dependence) represents a frequency-domain quantity with RMS magnitude, and the raised asterisk indicates a complex conjugate. The overbar indicates an average over time or over an ensemble of samples, as appropriate. As (5.1) shows, $\overline{|i|^2(t)}$ can be defined in either the frequency or time domain and, in the frequency domain, as a magnitude square quantity. Since we have assumed that $\Delta f = 1$, it will not be stated explicitly, unless necessary, in the future.

Voltage noise sources are described in the same manner:

$$\overline{|v|^2(t)} = \overline{vv^*}\Delta f = \overline{|v|^2}\Delta f = \lim_{T \to \infty} \frac{1}{T}\int_0^T v^2(t)dt \tag{5.2}$$

where the quantities are defined in the same manner as the currents. Many noise sources, most notably thermal sources, can be expressed as either noise currents or noise voltages. Because of their physical nature, many noise sources must be expressed as currents, but only rarely do we encounter a noise source that must be expressed as a voltage. Therefore, in most cases, we shall deal with noise currents, and include voltages only when necessary. In any case, as we shall see, it is always possible to convert one representation to the other.

Most electronic circuits include multiple noise sources. In general, at least a few of these are correlated; that is, the correlation, $C_{i;j,\,k}$, is nonzero:

$$C_{i;j,\,k} = \overline{i_j i_k^*} \tag{5.3}$$

where j and k are the indices of some pair of noise-current sources. In some cases, we encounter correlated voltage and current noise sources; the correlation is defined identically:

$$C_{a;j,k} = \overline{i_j v_k^*} \quad (5.4)$$

The general subject of correlation in the frequency domain is discussed in Section 2.2.3.

5.2 TWO-PORT NOISE ANALYSIS

The most common problem we encounter is noise in two-ports. Many electronic components, especially transistors, can be modeled as two-ports, so the treatment of two-port noise is especially important. From the noise parameters of a two-port, it is relatively easy to generate a correlation matrix. The correlation matrix can then be used directly in computer-aided circuit design.

A noisy two-port can be modeled as a noiseless two-port and a pair of noise sources. Those sources can be connected in a number of ways; the simplest is a current source in shunt with each port. The sources are, in general, correlated, so four parameters—the two mean-square currents and the real and imaginary parts of the correlations—are necessary to characterize the noise. Other representations are possible and have their own advantages and disadvantages, but in all cases, four parameters provide a full description of the noise at any particular frequency.

5.2.1 Two-Port Noise Representations

We consider first a two-port described by its admittance matrix. The noise at each port can simply be added to the equations describing the port voltage-current characteristics:

$$\begin{aligned} I_1 &= Y_{1,1}V_1 + Y_{1,2}V_2 + i_1 \\ I_2 &= Y_{2,1}V_1 + Y_{2,2}V_2 + i_2 \end{aligned} \quad (5.5)$$

These relations describe the circuit shown in Figure 5.1(a), a two-port having noise sources in shunt with each port. The noise sources, i_1 and i_2, are the short-circuit noise currents at the ports; when $V_1 = V_2 = 0$,

$$\begin{aligned} I_1 &= i_1 \\ I_2 &= i_2 \end{aligned} \quad (5.6)$$

A second possibility is to use an impedance matrix. Then

$$\begin{aligned} V_1 &= Z_{1,1}I_1 + Z_{1,2}I_2 + v_1 \\ V_2 &= Z_{2,1}I_1 + Z_{2,2}I_2 + v_2 \end{aligned} \tag{5.7}$$

and, as shown in Figure 5.1(b), the two-port noise is modeled by series voltage noise sources at each port, and each of these sources represents the open-circuit noise voltage at the port's respective terminals. Finally, we could use a hybrid of the two:

$$\begin{aligned} V_1 &= Z_{1,1}I_1 + Z_{1,2}I_2 + v \\ I_1 &= Y_{1,1}V_1 + Y_{1,2}V_2 + i \end{aligned} \tag{5.8}$$

This last formulation is clumsy to use, as it cannot be put into matrix form. It is better to use the chain (ABCD) matrix. Then

$$\begin{aligned} V_1 &= AV_2 + B(-I_2) + v \\ I_1 &= CV_2 + D(-I_2) + i \end{aligned} \tag{5.9}$$

The noise voltage and current are the same in either case. The use of (5.9), however, facilitates matrix calculations.

This last case is especially interesting, because it creates two noise sources, a voltage source and a current source, that are both located at the input of the two-port. This case is illustrated in Figure 5.1(c). In the first two cases, a noise analysis requires knowledge of the two-port's Y or Z parameters; in the third, two-port parameters are not needed. This recognition allows us to use the noise sources alone, in the absence of the two-port, to optimize noise figure.

It is relatively easy to transform these representations. For example, suppose we wish to transform the hybrid representation in Figure 5.1(c) to the admittance representation in Figure 5.1(a). We first observe that the short-circuit noise currents, i_1 and i_2, are given in Figure 5.1(c) by

$$\begin{aligned} i_1 &= i - Y_{1,1}v \\ i_2 &= -Y_{2,1}v \end{aligned} \tag{5.10}$$

or, in matrix form,

$$\begin{bmatrix} i_1 \\ i_2 \end{bmatrix} = \begin{bmatrix} -Y_{1,1} & 1 \\ -Y_{2,1} & 0 \end{bmatrix} \begin{bmatrix} v \\ i \end{bmatrix} \tag{5.11}$$

This can be expressed in matrix notation as

$$\mathbf{I} = \mathbf{TA} \tag{5.12}$$

where $\mathbf{T}$ is the matrix in (5.11), which we call the *transformation matrix*, and $\mathbf{A}$ is the vector containing v and i. The correlation matrix of the admittance-form sources is

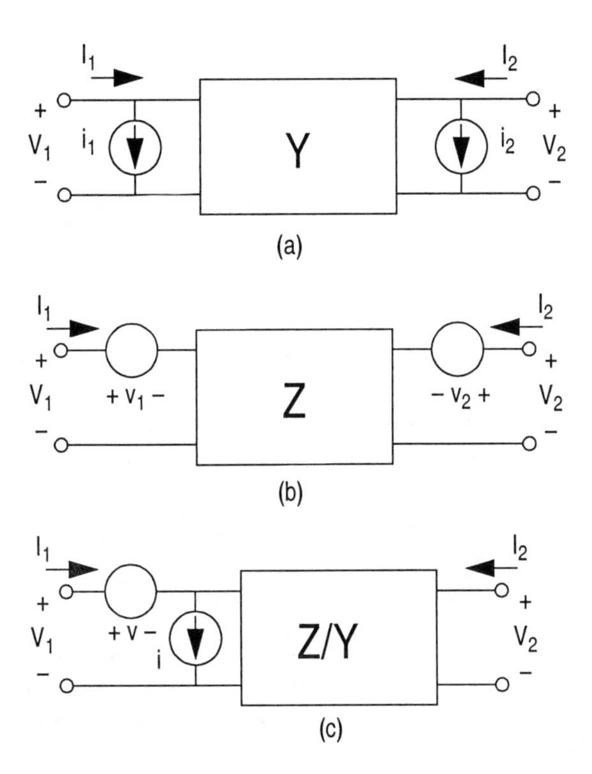

Figure 5.1 Noise equivalent circuits of two-ports: (a) admittance-matrix representation; (b) impedance-matrix representation; (c) hybrid impedance/admittance representation.

$$\mathbf{C}_i = \begin{bmatrix} \overline{i_1 i_1^*} & \overline{i_1 i_2^*} \\ \overline{i_2 i_1^*} & \overline{i_2 i_2^*} \end{bmatrix} = \overline{\mathbf{II}^{*\mathrm{T}}} \tag{5.13}$$

where the superscript * indicates a complex conjugate and the superscript T indicates the transpose. Substituting (5.12) into the right side of (5.13) and applying a little vector algebra gives

$$\mathbf{C}_i = \mathbf{T}\mathbf{C}_a\mathbf{T}^{*\mathrm{T}} \tag{5.14}$$

For other transformations we use (5.14) with a different transformation matrix. Equation (5.10) can be manipulated to create transformations between all three forms of the two-port noise model. The results are shown in Table 5.1, in which the column headings show the original representation and the rows give the resulting one.

Table 5.1 Noise-Source Transformations

Representation	Admittance	Impedance	Chain
Admittance	$\begin{bmatrix} 1 & 0 \\ 0 & 1 \end{bmatrix}$	$\begin{bmatrix} Y_{1,1} & Y_{1,2} \\ Y_{2,1} & Y_{2,2} \end{bmatrix}$	$\begin{bmatrix} -Y_{1,1} & 1 \\ -Y_{2,1} & 0 \end{bmatrix}$
Impedance	$\begin{bmatrix} Z_{1,1} & Z_{1,2} \\ Z_{2,1} & Z_{2,2} \end{bmatrix}$	$\begin{bmatrix} 1 & 0 \\ 0 & 1 \end{bmatrix}$	$\begin{bmatrix} 1 & -Z_{1,1} \\ 0 & -Z_{2,1} \end{bmatrix}$
Chain	$\begin{bmatrix} 0 & B \\ 1 & D \end{bmatrix}$	$\begin{bmatrix} 1 & -A \\ 0 & -C \end{bmatrix}$	$\begin{bmatrix} 1 & 0 \\ 0 & 1 \end{bmatrix}$

Columns: original representation; rows: resulting representation.

5.2.2 Noise Two-Port

We now focus on the noise model of Figure 5.1(c). We note that the two-port, being noiseless, has no effect on the signal-to-noise ratio once our signal reaches its input terminals. We can therefore dispense with the two-port and examine the noise circuit alone, including, of course, its source impedance or admittance. This is shown in Figure 5.2(a).

In general, the noise sources v and i in Figure 5.2(a) are correlated. It would be helpful to remove the correlation. To do this, we recognize that either of a pair of correlated quantities can be represented by a part that is perfectly correlated and another that is uncorrelated. Thus, we have

$$i = i_u + vY_{\text{cor}} \tag{5.15}$$

or

$$v = v_u + iZ_{\text{cor}} \tag{5.16}$$

where the quantities with the subscript u are the uncorrelated components, and Y_{cor} and Z_{cor} are called the *correlation admittance* and *correlation impedance*, respectively, where

$$\begin{aligned} Y_{\text{cor}} &= \Gamma\sqrt{\frac{\overline{|i|^2}}{\overline{|v|^2}}} = \frac{\overline{iv^*}}{\overline{|v|^2}} \\ Z_{\text{cor}} &= \Gamma\sqrt{\frac{\overline{|v|^2}}{\overline{|i|^2}}} = \frac{\overline{vi^*}}{\overline{|i|^2}} \end{aligned} \tag{5.17}$$

and Γ is the correlation coefficient between v and i. This formulation gives rise to the representations in Figure 5.2(b) and (c), in which Y_{cor} and Z_{cor} are treated as a physical admittance and impedance. It is important to recognize, in this formulation, that the real parts of Y_{cor} and Z_{cor} are noiseless.

In Section 2.3.1 we noted that a noise source can be represented by an equivalent noise resistance or conductance at the standard temperature of 290K. This is a common practice, even when the source of the noise is non-thermal; in effect, we use the resistance or conductance simply to describe the noise spectral density. Then, from (2.50) and (2.51), we define

$$G_n = \frac{\overline{|i_u|^2}}{4KT_0\Delta f} \qquad g_n = \frac{\overline{|i|^2}}{4KT_0\Delta f}$$
$$R_n = \frac{\overline{|v|^2}}{4KT_0\Delta f} \qquad r_n = \frac{\overline{|v_u|^2}}{4KT_0\Delta f} \tag{5.18}$$

Either representation—Figure 5.2(b) or (c)—could be used in the following derivation. As there is no need to do both, we shall more or less arbitrarily

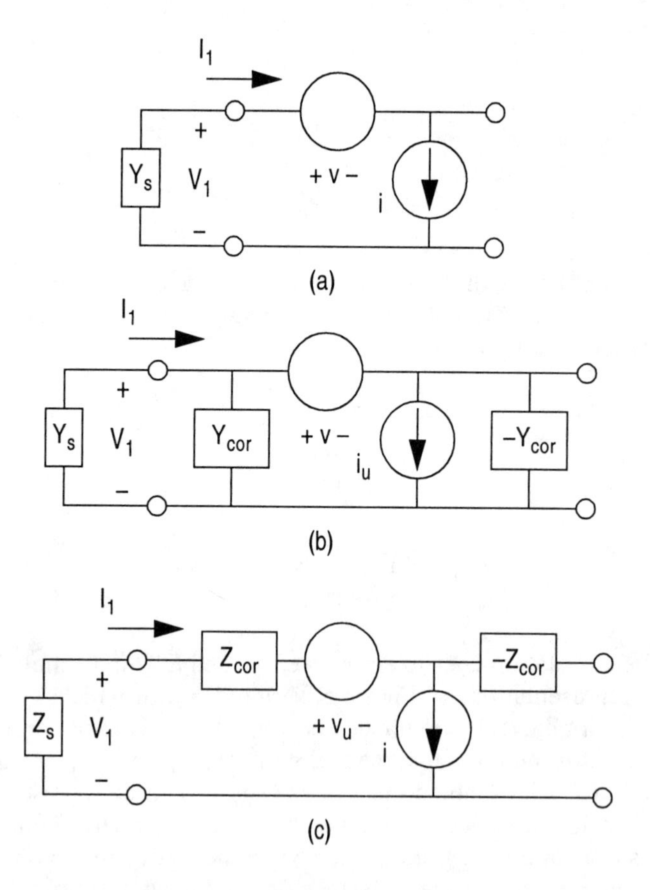

Figure 5.2 Representation of the noise two-port using correlated sources (a) and two dual cases, (b) and (c), in which the noise of the sources is uncorrelated.

use the admittance form in Figure 5.2(b). The quantities are, in any case, easily converted. From (5.15) to (5.18) we obtain

$$g_n = G_n + R_n|Y_{cor}|^2 \qquad R_n = r_n + g_n|Z_{cor}|^2 \tag{5.19}$$

$$r_n = \frac{G_n}{|Y_{cor}|^2 + G_n/R_n} \qquad G_n = \frac{r_n}{|Z_{cor}|^2 + r_n/g_n} \tag{5.20}$$

and

$$Z_{cor} = \frac{Y^*_{cor}}{|Y_{cor}|^2 + G_n/R_n} \qquad Y_{cor} = \frac{Z^*_{cor}}{|Z_{cor}|^2 + r_n/g_n} \tag{5.21}$$

5.2.3 Noise Figure

We now can determine the noise figure of the two-port in Figure 5.1(c). The noise figure clearly must be a function of the source impedance or admittance, which we shall designate Z_s or Y_s, respectively. According to the definition given in Section 3.1.3, the noise figure of the circuit, F, is

$$F = \frac{\overline{|i_{tot}|^2}}{\overline{|i_s|^2}} \tag{5.22}$$

where $\overline{|i_{tot}|^2}$ is the total short-circuit mean-square noise current at the rightmost terminals of the circuit in Figure 5.2(b) or (c), and $\overline{|i_s|^2}$ is that portion of the mean-square current arising from Z_s or Y_s only. Since the noise sources contributing to these quantities are uncorrelated, it is a straightforward matter to calculate them. The total short-circuit current in terms of the noise conductances is

$$\overline{|i_{tot}|^2}/4KT_0\Delta f = G_s + G_n + R_n|Y_s + Y_{cor}|^2 \tag{5.23}$$

where $G_s = \text{Re}\{Y_s\}$. The portion of the short-circuit noise created by the source is simply

$$\overline{|i_s|^2}/4KT_0\Delta f = G_s \tag{5.24}$$

The noise figure becomes

$$F = 1 + \frac{1}{G_s}(G_n + R_n|Y_s + Y_{\text{cor}}|^2) \tag{5.25}$$

Minimizing F requires, among other things, $\text{Im}\{Y_s\} = -\text{Im}\{Y_{\text{cor}}\}$. Applying this constraint and differentiating gives the minimum noise figure, F_{min}:

$$\begin{aligned} F_{\text{min}} &= 1 + 2R_n(G_{\text{cor}} + G_{s,\,\text{opt}}) \\ &= 1 + 2(R_n G_{\text{cor}} + \sqrt{R_n G_n + (R_n G_{\text{cor}})^2}) \end{aligned} \tag{5.26}$$

where we have used the optimum value of G_s, $G_{s,\,\text{opt}}$:

$$G_s = G_{s,\,\text{opt}} = \sqrt{\frac{G_n}{R_n} + G_{\text{cor}}^2} \tag{5.27}$$

These expressions impose an important constraint on R_n, F_{min}, and $G_{s,\,\text{opt}}$. Solving (5.26) for G_{cor}, substituting into (5.27), and solving for G_n gives

$$G_n = (F_{\text{min}} - 1)\left(G_{s,\,\text{opt}} - \frac{F_{\text{min}} - 1}{4R_n}\right) \tag{5.28}$$

Since G_n represents the mean-square current of a noise source, we must have $G_n \geq 0$. This consideration creates the following constraint on noise parameters:

$$R_n \geq \frac{F_{\text{min}} - 1}{4G_{s,\,\text{opt}}} \tag{5.29}$$

With further algebraic manipulation, it is possible to remove the correlation admittance explicitly from (5.25). (It is still implicit in the optimum source admittance, $Y_{s,\,\text{opt}}$, and minimum noise figure.) We obtain

$$F = F_{\min} + \frac{R_n}{G_s}|Y_s - Y_{s,\,\mathrm{opt}}|^2 \tag{5.30}$$

This expression is the one most frequently cited to describe the noise figure of a two-port as a function of source admittance. The quantity R_n, having units of resistance, is normally treated simply as an empirically determined parameter of the two-port. It establishes the sensitivity of the noise figure to variations of the source admittance.

The four noise parameters of the two-port—$F_{\min}$, R_n, and the real and imaginary parts of $Y_{s,\,\mathrm{opt}}$—are a complete characterization of the two-port at any particular frequency. Once these parameters are measured, a noise correlation matrix for the two-port can be generated and that matrix can be combined with the correlation matrix of a much larger circuit. The measurement of these four noise parameters is, of course, of paramount importance. The obvious method is to measure F at a number of values of Y_s and to fit the four parameters in a least-squares sense to the measured results. In the early days of noise research, such methods were not possible, so other techniques were developed. Today, with advances in automated measurement, that method is now thoroughly practical and is the dominant one for determining transistor noise parameters.

5.2.4 Correlation Matrix and Noise Parameters

In computer-aided circuit design, we usually need the admittance-form correlation matrix of the noisy two-port. It is easily found from the noise parameters of (5.30), $F_{\min}$, R_n, and $Y_{s,\,\mathrm{opt}}$. We first obtain G_n from (5.28) and then determine Y_{cor} from $Y_{s,\,\mathrm{opt}}$. From (5.27),

$$G_{\mathrm{cor}} = \sqrt{G_{s,\,\mathrm{opt}}^2 - \frac{G_n}{R_n}} \tag{5.31}$$

and, as before, $\mathrm{Im}\{Y_{\mathrm{cor}}\} = -\mathrm{Im}\{Y_{s,\,\mathrm{opt}}\}$. We find, with a little algebra, that

$$\mathbf{C}_i = \begin{bmatrix} G_n + |Y_{1,1} - Y_{\mathrm{cor}}|^2 R_n & Y_{2,1}^*(Y_{1,1} - Y_{\mathrm{cor}})R_n \\ Y_{2,1}(Y_{1,1} - Y_{\mathrm{cor}})^* R_n & |Y_{2,1}|^2 R_n \end{bmatrix} \tag{5.32}$$

Inverting these equations gives the noise parameters as a function of the correlation-matrix elements. We obtain

$$R_n = \frac{C_{i;2,2}}{|Y_{2,1}|^2} \tag{5.33}$$

$$Y_{cor} = Y_{1,1} - \frac{C_{i;1,2}}{C_{i;2,2}} Y_{2,1} \tag{5.34}$$

G_n is found from

$$G_n = C_{i;1,1} - |Y_{1,1} - Y_{cor}|^2 R_n \tag{5.35}$$

then F_{min} is found from (5.26) and $Y_{s,\,opt}$ from (5.27).

One must be careful in applying these conversions to noise parameters of transistors, as R_n is often provided on transistor data sheets in normalized form. That is, $\hat{R}_n$ is given,[1] where

$$\hat{R}_n = \frac{R_n}{Z_0} \tag{5.36}$$

and Z_0 is the reference impedance for the transistor's scattering parameters, usually 50Ω.

5.2.5 Noise Circles

In this section, we examine the remarkable fact that contours of constant noise figure are circles in the plane of source reflection coefficient. We begin by noting that (5.30) can be written in normalized form as

$$F = F_{min} + \frac{\hat{R}_n}{\hat{G}_s} |\hat{Y}_s - \hat{Y}_{s,\,opt}|^2 \tag{5.37}$$

where

---

1. Unfortunately, in many transistor data sheets and other publications, the normalized noise resistance is called r_n, the same variable used as a noise resistance in Section 5.2.2. It should be clear that the quantity in (5.36) is very different from r_n in (5.18).

$$\hat{Y}_s = \frac{Y_s}{Y_0} \tag{5.38}$$

$Y_0 = 1/Z_0$ is the normalizing admittance, $\hat{G}_s = \mathrm{Re}\{\hat{Y}_s\}$, and $\hat{R}_n$ is given by (5.36). The normalized source admittance, $\hat{Y}_s$, in terms of source reflection coefficient, is

$$\hat{Y}_s = \frac{1 - \Gamma_s}{1 + \Gamma_s} \tag{5.39}$$

and the optimum source admittance, $\hat{Y}_{s,\,\mathrm{opt}}$, is

$$\hat{Y}_{s,\,\mathrm{opt}} = \frac{1 - \Gamma_{s,\,\mathrm{opt}}}{1 + \Gamma_{s,\,\mathrm{opt}}} \tag{5.40}$$

Substituting these into (5.37) gives the equivalent expression in terms of reflection coefficient,

$$F = F_{\min} + 4\hat{R}_n \frac{|\Gamma_s - \Gamma_{s,\,\mathrm{opt}}|^2}{(1 - |\Gamma_s|^2)|1 + \Gamma_{s,\,\mathrm{opt}}|^2} \tag{5.41}$$

To derive an expression for the noise-figure circles, we first move the constant quantities to one side of the equation and, for simplicity, call this N:

$$\frac{|\Gamma_s - \Gamma_{s,\,\mathrm{opt}}|^2}{1 - |\Gamma_s|^2} = N = \frac{F - F_{\min}}{4\hat{R}_n}|1 + \Gamma_{s,\,\mathrm{opt}}|^2 \tag{5.42}$$

The left side of the equation can then be manipulated algebraically to form

$$\left|\Gamma_s - \frac{\Gamma_{s,\,\mathrm{opt}}}{1 + N}\right|^2 = \frac{N^2 + N(1 - |\Gamma_{s,\,\mathrm{opt}}|^2)}{(1 + N)^2} \tag{5.43}$$

This expression represents a circle in the plane of source reflection coefficient. The center, c, of each circle is

$$c = \frac{\Gamma_{s,\,\mathrm{opt}}}{1 + N} \tag{5.44}$$

and the radius, r, is

$$r = \frac{1}{1 + N}\sqrt{N^2 + N(1 - |\Gamma_{s,\,\mathrm{opt}}|^2)} \tag{5.45}$$

The values of N, on which the circles are parameterized, represent values of noise figure.

Figure 5.3 shows an example of noise circles of a MESFET at 8 GHz drawn at 1-dB increments. The parameters, taken from a data sheet, are $F_{\mathrm{min}} = 1.35$ dB, $\hat{R}_n = 0.40$ (normalized to 50Ω), and $\Gamma_{\mathrm{opt}} = 0.5\angle 152°$. The centers of all the circles are on a line in the S plane, as indicated by (5.44); their centers move closer to the origin, and the circle radius increases, with F.

5.3 NOISE ANALYSIS OF LARGE CIRCUITS

Frequently we are faced with the problem of analyzing noise in large circuits, often by computer-aided techniques. Large circuits invariably contain multiple noise sources, many—but not all—of which may be correlated. The circuit is effectively a multiport, in which the terminals of each noise source represent a port.

Inevitably we have two tasks. The first is to formulate the circuit equations in a manner that is amenable to noise analysis; the second is to add the noise information to the circuit equations and to determine noise voltages at the nodes, noise figure, or noise temperature. It happens that the most convenient form of the circuit equations is a nodal formulation, and the admittance-form description of the noisy elements is most amenable to the nodal formulation.

5.3.1 Nodal Circuit Equations

The circuit equations of high-frequency circuits can be formulated in a number of different ways that are amenable to computer manipulation. After much research on this subject throughout the latter half of the twentieth century, a consensus emerged that a nodal formulation is most practical.

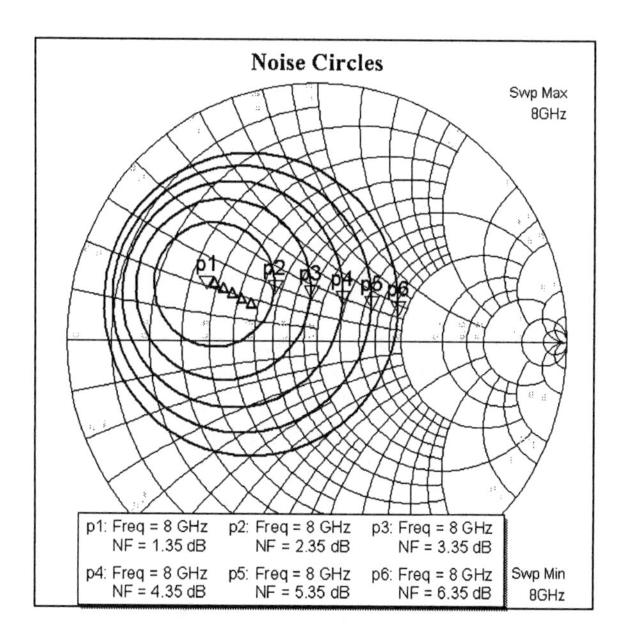

Figure 5.3 Example of a set of noise circles of a MESFET at 8 GHz. The minimum noise figure is 1.35 dB and contours are plotted at 1-dB increments. The plane represents source reflection coefficient.

The nodal formulation is based on the concept of an *indefinite admittance matrix*. The matrix itself is not particularly useful, but it can be created easily by computer and manipulated to form a multiport or multinode admittance matrix. Once the matrix is formulated and reduced, we call it the *nodal matrix* of the circuit.

5.3.1.1 Formulating the Matrix

The idea behind the indefinite admittance matrix is illustrated in Figure 5.4. The terminals represent the nodes of the circuit, and the terminal voltages are referenced to an arbitrary ground potential. We can write a set of admittance equations for this network:

$$\begin{bmatrix} I_1 \\ I_2 \\ \cdots \\ I_N \end{bmatrix} = \begin{bmatrix} Y_{1,1} & Y_{1,2} & \cdots & Y_{1,N} \\ Y_{2,1} & Y_{2,2} & \cdots & Y_{2,N} \\ \cdots & \cdots & \cdots & \cdots \\ Y_{N,1} & Y_{N,2} & \cdots & Y_{N,N} \end{bmatrix} \begin{bmatrix} V_1 \\ V_2 \\ \cdots \\ V_N \end{bmatrix} \tag{5.46}$$

This matrix has a number of important properties. Since the network is floating, the sum of all the nodal currents must be zero. This means, in turn, that the columns of the matrix must sum to zero. Although it is less obvious intuitively, the rows must sum to zero as well.

One of the nicest properties of the indefinite admittance matrix is the ease with which it can be formulated. Suppose we connect a two-terminal element of admittance Y_e between nodes j and k. This increases the node currents as

$$\begin{aligned} \Delta I_j &= Y_e(V_j - V_k) \\ \Delta I_k &= Y_e(V_k - V_j) \end{aligned} \tag{5.47}$$

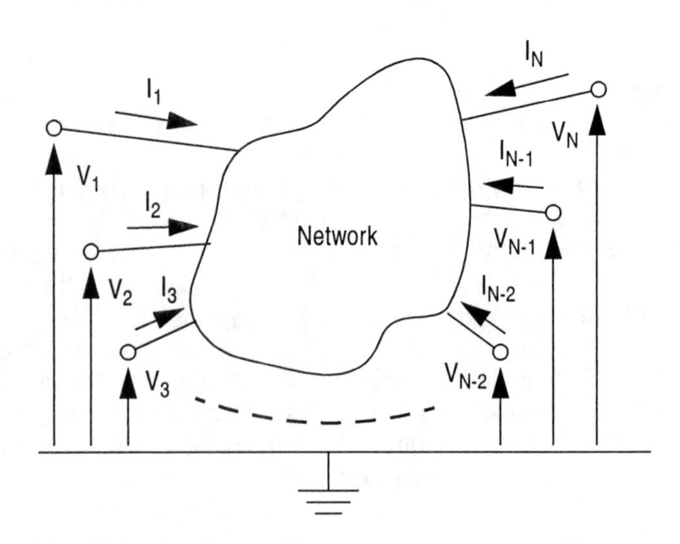

Figure 5.4 Illustration of a circuit represented by an indefinite admittance matrix. All node voltages are referenced to an arbitrary ground potential, and no terminals of the network are connected to ground.

which tells us to add Y_e to Y_{jj} and Y_{kk} and to subtract it from Y_{jk} and Y_{kj}. If either the jth or kth row-column pair do not exist in the matrix, the matrix is simply augmented by that row-column pair, and the admittances are added to the new row and column. Note that this process does not change the zero-sum property of the rows and columns.

This pattern of adding and subtracting a value from a set of particular matrix locations is sometimes called a *stamp*, and adding the element to the matrix is called *stamping the matrix*. Indeed, the process of adding element values to circuit equations is as mindless as the term implies, perfect for computers. It involves stamping the matrix with the appropriate stamp, adding new row-column pairs as needed.

Similar stamps can be derived for other types of elements. For example, a voltage-controlled current source has

$$\begin{aligned} I_p &= A(V_j - V_k) \\ I_q &= A(V_k - V_j) \end{aligned} \tag{5.48}$$

where (p, q) are the current-source nodes and (j, k) are the control-voltage nodes. A, the transconductance, is added to the (p, j) and (q, k) positions, while $-A$ is added to (p, k) and (q, j).

In general, any admittance element can be added to the matrix by an application of the above procedure. Suppose we are given the admittance matrix of an n-port element and must add this to the circuit matrix. Figure 5.5 illustrates the problem. The n-port element's port no. 1 is realized by

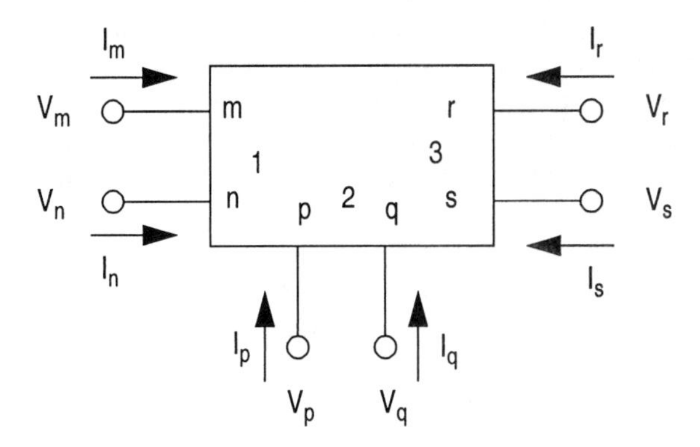

Figure 5.5 Port matrix for the example in Section 5.3.1.1. It is converted to a nodal matrix before it is combined with the circuit matrix.

node pairs (m, n), port 2 by (p, q), port 3 by (r, s), and so on. For the n-port, we have

$$\begin{aligned} I_m &= Y_{1,1}(V_m - V_n) + Y_{1,2}(V_p - V_q) + Y_{1,3}(V_r - V_s) + \dots \\ I_n &= -Y_{1,1}(V_m - V_n) - Y_{1,2}(V_p - V_q) - Y_{1,3}(V_r - V_s) - \dots \end{aligned} \tag{5.49}$$

and so on for the other port currents. This can be expressed in the form

$$\begin{bmatrix} I_m \\ I_n \\ I_p \\ I_q \\ \dots \end{bmatrix} = \begin{bmatrix} Y_{1,1} & -Y_{1,1} & Y_{1,2} & -Y_{1,2} & \dots \\ -Y_{1,1} & Y_{1,1} & -Y_{1,2} & Y_{1,2} & \dots \\ Y_{2,1} & -Y_{2,1} & Y_{2,2} & -Y_{2,2} & \dots \\ -Y_{2,1} & Y_{2,1} & -Y_{2,2} & Y_{2,2} & \dots \\ \dots & \dots & \dots & \dots & \dots \end{bmatrix} \begin{bmatrix} V_m \\ V_n \\ V_p \\ V_q \\ \dots \end{bmatrix} \tag{5.50}$$

which now has the structure of an indefinite admittance matrix. To add an element's indefinite admittance matrix to the circuit matrix, we simply add the element's matrix to the corresponding positions in the circuit matrix.

5.3.1.2 Matrix Reduction

We now must reduce the matrix to a conventional one, either a nodal matrix or an n-port matrix. At present, the matrix voltages and currents represent all the nodes, not just the ones we are interested in; also, disturbingly, it is singular, so we are severely limited in our ability to manipulate it.

The singularity is caused by the fact that the node voltages are indistinct. For example, we could add a quantity to all the node voltages on the right side of (5.46) and the equations would still be satisfied. To remove the singularity, we could ground one node, something we usually wish to do in any case. Grounding the kth node sets V_k to zero, so V_k is eliminated from the right-side vector and the kth column, which is multiplied by V_k, is eliminated from the matrix. The kth current component, I_k, is then no longer of interest, so the kth row and I_k are removed as well. This gives us a square, nonsingular matrix.[2]

Generally, we are interested in the voltages and currents at a few nodes, usually ones to which we make connections. We do not need to know the

2. Of course, the matrix could have other problems which might lead to singularity. Any such problems would have to be fixed as well.

voltages at the others, and we already know that their currents are zero, since they are never connected to anything. We would like to reduce the size of the matrix accordingly. One possible method is to invert the matrix, forming an impedance matrix,

$$\begin{bmatrix} V_1 \\ V_2 \\ \dots \\ V_N \end{bmatrix} = \begin{bmatrix} Z_{1,1} & Z_{1,2} & \dots & Z_{1,N} \\ Z_{2,1} & Z_{2,2} & \dots & Z_{2,N} \\ \dots & \dots & \dots & \dots \\ Z_{N,1} & Z_{N,2} & \dots & Z_{N,N} \end{bmatrix} \begin{bmatrix} I_1 \\ I_2 \\ \dots \\ I_N \end{bmatrix} \tag{5.51}$$

and to set the currents of the unwanted nodes to zero. The columns corresponding to those nodes are then eliminated, as well as the rows (since the voltages, while valid, are not of interest). This reduced impedance matrix can then be converted to admittance parameters, scattering parameters, or any other desired form.

The above process, while technically correct, is numerically inefficient. A better method is as follows:

1. Factor the admittance matrix by LU decomposition. (See any good book on numerical methods for further information on this process.)
2. Set the current for one "desired" node $I_n = 1$; set all other currents equal to zero.
3. Back-substitute to obtain the node-voltage vector; these voltages are numerically equal to $Z_{j,\,n}$, the column of the impedance matrix (5.51) corresponding to node n.
4. Discard the elements of $Z_{j,\,n}$ that correspond to unwanted nodes.
5. We now have the column of the reduced impedance matrix corresponding to node n. Repeat this process for all other nodes of interest.
6. We now have a reduced impedance matrix. It can be inverted to form a reduced admittance matrix if desired.

This method reduces the numerical effort by eliminating back-substitution for unwanted nodes.

It is important to note that the unreduced indefinite admittance matrix is extremely sparse. The number of entries in any row or column is equal to the number of elements connected to the node that it represents. A large circuit might have thousands of nodes, but only a few elements connected to any particular node, so the matrix has mostly zero entries. It is wasteful to use memory to store large numbers of zeros, and to spend computational effort multiplying zero by zero and adding it to zero.

For this reason, sparse-matrix solvers, which avoid storing and manipulating zero entries, have been developed. Such methods can speed the factoring of the matrix, especially as the matrix becomes larger. Classical Gaussian reduction is an N^3 process, where N is the dimension of the matrix; that is, the computational effort increases as the cube of the matrix's size. Sparse-matrix techniques can reduce this to $\sim N^{1.5}$. For this reason, sparse-matrix methods are used in all modern circuit simulators.

5.3.2 Noise Analysis

5.3.2.1 Noise Sources and Correlations

Invariably, a large circuit contains a large number of noise sources. These sources represent the noise generated in solid-state devices and thermal noise of passive, resistive circuit elements. As with two-ports, we make no assumptions about the physical source of the noise (although occasionally it may be helpful in creating the noise-correlation matrix for the circuit); we assume it only to be white, Gaussian noise over the bandwidth of interest.

In general, a vector of sources, **I**, represents the set of noise sources in the circuit:

$$\mathbf{I} = \begin{bmatrix} i_1 \\ i_2 \\ \cdots \\ i_N \end{bmatrix} \tag{5.52}$$

The sources and their correlations are defined by a correlation matrix, $\mathbf{C}_i$:

$$\mathbf{C}_i = \overline{\mathbf{I}\mathbf{I}^{*\mathrm{T}}} = \begin{bmatrix} \overline{i_1 i_1^*} & \overline{i_1 i_2^*} & \dots & \overline{i_1 i_N^*} \\ \overline{i_2 i_1^*} & \overline{i_2 i_2^*} & \dots & \overline{i_2 i_N^*} \\ \dots & \dots & \dots & \dots \\ \overline{i_N i_1^*} & \overline{i_N i_2^*} & \dots & \overline{i_N i_N^*} \end{bmatrix} \tag{5.53}$$

where the raised T indicates the transpose of the vector. From (5.1), the terms on the diagonal are mean-square currents. The off-diagonal terms represent cross correlations; these are zero if the respective sources are uncorrelated. Sometimes, instead of cross correlations, correlation coefficients $\Gamma_{j,\,k}$ are used:

$$\Gamma_{j,\,k} = \frac{\overline{i_j i_k^*}}{\sqrt{\overline{|i_j|^2}\,\overline{|i_k|^2}}} \tag{5.54}$$

Note that $\mathbf{C}_i$ could have been defined as

$$\hat{\mathbf{C}}_i = \overline{\mathbf{I}^*\mathbf{I}^{\mathrm{T}}} = \begin{bmatrix} \overline{i_1^* i_1} & \overline{i_1^* i_2} & \dots & \overline{i_1^* i_N} \\ \overline{i_2^* i_1} & \overline{i_2^* i_2} & \dots & \overline{i_2^* i_N} \\ \dots & \dots & \dots & \dots \\ \overline{i_N^* i_1} & \overline{i_N^* i_2} & \dots & \overline{i_N^* i_N} \end{bmatrix} = \mathbf{C}_i^* \tag{5.55}$$

A brief review of the literature shows that the representations in (5.53) and (5.55) are used almost equally; there is no de facto standard for the correlation-matrix representation. In this book, our standard is the form of (5.53). This distinction is obviously important, as (5.53) and (5.55) are complex conjugates. The reader should be mindful of this distinction in reading other literature, which might use a different representation from ours.

5.3.2.2 Correlation Matrix of a Passive Circuit

Usually the noise correlation matrix of a component must be measured in some manner. For example, the noise parameters of a transistor, $F_{\min}$, R_n, and $Y_{s,\,opt}$, described in Section 5.2.3, must be measured, and the noise correlation matrix is then derived from them. If a circuit consists only of passive elements, however, the noise correlation matrix can be derived directly from the admittance parameters [5.4]. The matrix is simply

$$\mathbf{C}_i = 4KT\Delta f\,\mathrm{Re}\{\mathbf{Y}\} \tag{5.56}$$

where **Y** is the admittance matrix of the passive circuit. For (5.56) to be valid, the circuit must consist of passive elements only, and all lossy elements must generate ordinary thermal noise. However, the circuit need not be reciprocal.

It is important to note that, in (5.56), T is the physical temperature in Kelvins, not the standard noise temperature, $T_0 = 290\mathrm{K}$. To maintain good matrix conditioning in the correlation matrix, it is a common practice to make $\Delta f = 1$ and to normalize the matrix to $4KT_0$. In this case the entries should be

$$\mathbf{C}_i = \frac{T}{T_0}\mathrm{Re}\{\mathbf{Y}\} \tag{5.57}$$

that is, the entries are simply scaled according to their absolute temperature.

This relation applies to two-terminal elements as well, in which **Y** in (5.57) becomes a simple scalar. In that case, (5.50) shows that the resulting $\mathbf{C}_i$ is simply a matrix of dimension 2. Equation (5.57) also shows that, to treat a lossy element as noiseless, we simply neglect to include it in the circuit's correlation matrix.

5.3.2.3 Formulation of the Nodal Noise Correlation Matrix

We now consider the problem of integrating the noise correlation matrix with the nodal or port admittance matrix generated as described in Section 5.3.1. The admittance matrix may have either a nodal or port formulation; in this section, we assume that a nodal formulation is used. There are two reasons for this approach: first, the nodal formulation is more general, and second, the use of a port formulation is a straightforward variation of the nodal description.

The circuit's nodal matrix must be formulated so that the terminals are available for connecting all noise sources. The reduced circuit may have more nodes than this, so many of those nodes may have no sources connected. The nodes of the correlation matrix of the circuit must correspond to those of the circuit's nodal matrix, so both matrices must have the same dimension. Many of the rows and columns of the correlation matrix are therefore zero, making it singular. Although we usually try to avoid creating singular matrices, in this case the singularity creates no difficulties.

Figure 5.6 shows the circuit, which follows Figure 5.5 to some degree. It shows noise sources connected to certain nodes but none connected to others. Following (5.52), we write the currents as

$$\mathbf{I} = \begin{bmatrix} i_m \\ i_n \\ i_p \\ i_q \\ i_r \\ i_s \\ \ldots \end{bmatrix} = \begin{bmatrix} i_1 \\ -i_1 \\ i_2 \\ -i_2 \\ i_3 \\ -i_3 \\ \ldots \end{bmatrix} \tag{5.58}$$

From (5.53), the correlation matrix is

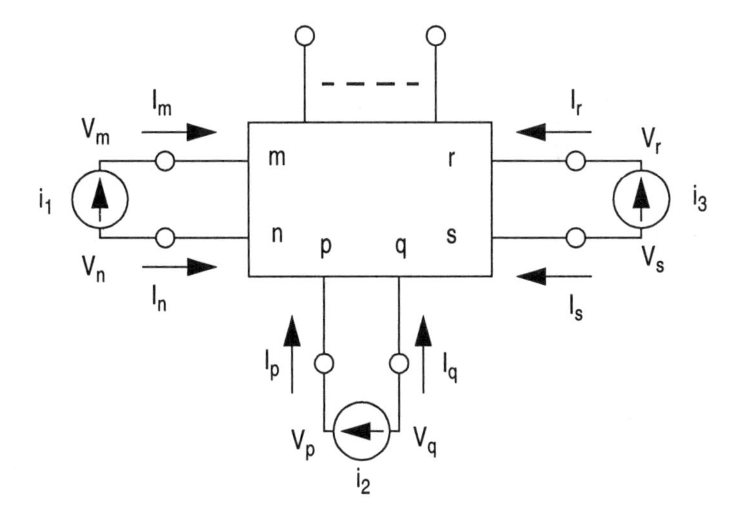

Figure 5.6 Connection of noise sources, possibly correlated, to the circuit. Some nodes of the circuit may not be connected to noise sources.

$$\mathbf{C}_i = \overline{\mathbf{I}\mathbf{I}^{*\mathrm{T}}} \tag{5.59}$$

Substituting (5.58) into (5.59) gives us the correlation matrix in nodal form, as a function of the port-current correlations:

$$\mathbf{C}_i = \begin{bmatrix} \overline{i_1 i_1^*} & -\overline{i_1 i_1^*} & \overline{i_1 i_2^*} & -\overline{i_1 i_2^*} & \dots \\ -\overline{i_1 i_1^*} & \overline{i_1 i_1^*} & -\overline{i_1 i_2^*} & \overline{i_1 i_2^*} & \dots \\ \overline{i_2 i_1^*} & -\overline{i_2 i_1^*} & \overline{i_2 i_2^*} & -\overline{i_2 i_2^*} & \dots \\ -\overline{i_2 i_1^*} & \overline{i_2 i_1^*} & -\overline{i_2 i_2^*} & \overline{i_2 i_2^*} & \dots \\ \dots & \dots & \dots & \dots & \dots \end{bmatrix} \tag{5.60}$$

Frequently we encounter the situation where the ports of a two-port have a common terminal. This occurs most frequently when a transistor, inherently a three-terminal element, is characterized as a two-port, and we need to generate a three-terminal representation from the two-port data. In this case, we need to generate a 3×3 correlation matrix. The problem is illustrated in Figure 5.7. We note that

$$\begin{aligned} i_m &= i_1 \\ i_n &= i_2 \\ i_p &= -(i_m + i_n) = -(i_1 + i_2) \end{aligned} \tag{5.61}$$

and substituting, as before, gives

$$\mathbf{C}_i = \begin{bmatrix} \overline{i_1 i_1^*} & \overline{i_1 i_2^*} & -\overline{i_1 i_1^*} - \overline{i_1 i_2^*} \\ \overline{i_2 i_1^*} & \overline{i_2 i_2^*} & -\overline{i_2 i_1^*} - \overline{i_2 i_2^*} \\ -\overline{i_1 i_1^*} - \overline{i_2 i_1^*} & -\overline{i_2 i_2^*} - \overline{i_1 i_2^*} & \begin{matrix} \overline{i_1 i_1^*} + \overline{i_2 i_1^*} \\ + \overline{i_1 i_2^*} + \overline{i_2 i_2^*} \end{matrix} \end{bmatrix} \tag{5.62}$$

5.3.2.4 Noise Analysis

Usually we wish to find one of a few possible quantities: the mean-square noise voltages at particular nodes, the mean-square noise voltage at the output port of the circuit (from which we easily obtain noise power), or noise parameters of the circuit when it is operated as a two-port. Whichever of these is our goal, we must first determine the noise voltage correlation matrix.

From Table 5.1, which is generally valid for multiports and multiterminal nodal matrices as well as two-ports, we have

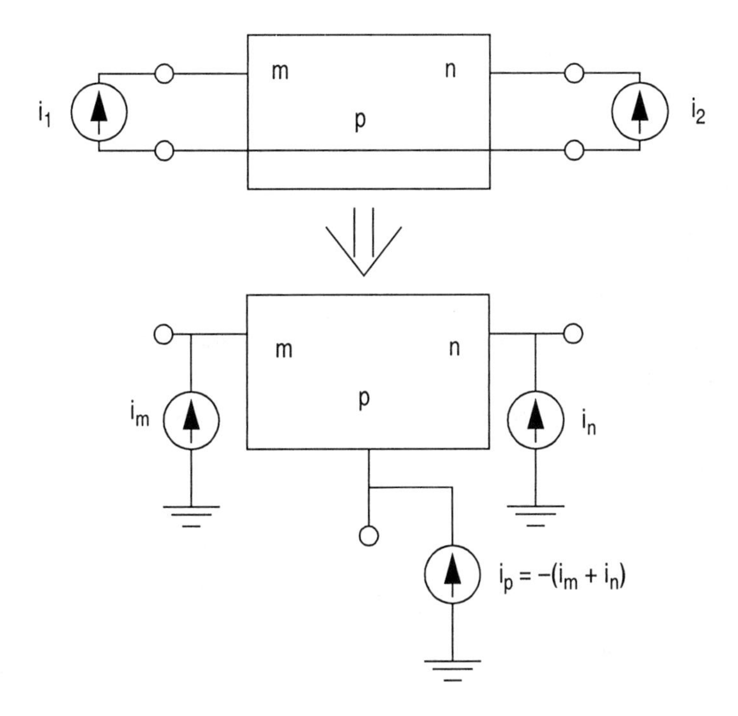

Figure 5.7 A two-port having a terminal common to both ports frequently must be converted to a three-port. This situation arises frequently when the grounded source of a FET or emitter of a bipolar transistor is "lifted" to allow for series feedback, use in a differential amplifier, or similar applications.

$$\mathbf{C}_i = \overline{\mathbf{I}\mathbf{I}^{*\mathrm{T}}} = \mathbf{Y}\mathbf{V}(\mathbf{Y}\mathbf{V})^{*\mathrm{T}} = \mathbf{Y}\mathbf{V}\mathbf{V}^{*\mathrm{T}}\mathbf{Y}^{*\mathrm{T}} = \mathbf{Y}\mathbf{C}_v\mathbf{Y}^{*\mathrm{T}} \tag{5.63}$$

or, equivalently,

$$\mathbf{C}_v = \mathbf{Z}\mathbf{C}_i\mathbf{Z}^{*\mathrm{T}} \tag{5.64}$$

where $\mathbf{C}_v$ is the voltage correlation matrix. In principle, $\mathbf{C}_v$ can be determined from (5.64) by inverting the admittance matrix to form the impedance matrix, and finally performing the multiplications. This approach is both inefficient and expensive in terms of both memory and computational effort. The following method is superior.

Let

$$\mathbf{M} = \mathbf{C}_v\mathbf{Y}^{*\mathrm{T}} \tag{5.65}$$

where $\mathbf{M}$ is an auxiliary matrix having the same dimensions as $\mathbf{Y}$ and $\mathbf{C}_v$. We first solve

$$\mathbf{Y}\mathbf{M} = \mathbf{C}_i \tag{5.66}$$

to obtain the matrix $\mathbf{M}$. This involves factoring $\mathbf{Y}$ by LU decomposition or other appropriate means and back-substituting to obtain the full $\mathbf{M}$ matrix. We need to retain, however, only the rows of $\mathbf{M}$ corresponding to nodes to which noise sources are connected.

Next, we solve

$$\mathbf{C}_v\mathbf{Y}^{*\mathrm{T}} = \mathbf{M} \tag{5.67}$$

This is best performed by solving the equivalent equation,

$$\mathbf{Y}\mathbf{C}_v^{*\mathrm{T}} = \mathbf{M}^{*\mathrm{T}} \tag{5.68}$$

This does not require further LU decomposition of $\mathbf{Y}$, which would be a computationally expensive operation. Furthermore, the retained rows of $\mathbf{M}$ are now columns of $\mathbf{M}^{*\mathrm{T}}$, so, if the matrix dimension is N, we perform the back-substitution at most N times. Finally, we transpose and conjugate $\mathbf{C}_v^{*\mathrm{T}}$ to obtain $\mathbf{C}_v$.

The mean-square noise voltage at any node is now the corresponding main-diagonal element of $\mathbf{C}_v$. To determine the output voltage, we need the individual components of $\mathbf{C}_v$ corresponding to the nodes of the port. If the node indices of the terminals of the output port are m and n, the mean-square voltage $\overline{|v_L|^2}$ is

$$\begin{aligned}\overline{|v_L|^2} &= \overline{|v_m - v_n|^2} = \overline{(v_m - v_n)(v_m^* - v_n^*)} \\ &= \overline{|v_m|^2} + \overline{|v_n|^2} - 2Re\{\overline{v_m v_n^*}\}\end{aligned} \tag{5.69}$$

Of course, in many high-frequency circuits, one node of the output port is ground. Then, if m is the output-port node,

$$\overline{|v_L|^2} = \overline{|v_m|^2} \tag{5.70}$$

Finally, the noise output power, P_L, is

$$P_L = \frac{\overline{|v_o|^2}}{Z_0} \tag{5.71}$$

where Z_0 is the load impedance, assumed to be real.

5.3.2.5 Noise Temperature, Noise Figure, and Noise Parameters

When the circuit has been reduced to a two-port, noise temperature is easily determined. Having found P_L (5.71), we calculate the equivalent input temperature, T_n, from (3.4):

$$T_n = \frac{P_L/K\Delta f}{G_t} \tag{5.72}$$

G_t is the transducer gain of the two-port, which presumably has been calculated as part of the linear circuit analysis. In the above formulation, the noise of the source and load terminations must not be included in the calculation of $\mathbf{C}_v$, but the terminations themselves must be included in the admittance matrix.

Noise figure, for the particular value of source impedance used to generate the nodal matrix, can be found by converting the noise temperature to a noise figure (3.10).

The reduced correlation matrix has everything needed to determine all noise parameters of the circuit. First, we must convert the nodal noise correlation matrix to a two-port matrix. Let us assume, for this purpose, that the nodes (m, n) are the input port and (p, q) are the output. Then

$$\begin{aligned} v_1 &= v_m - v_n \\ v_2 &= v_p - v_q \end{aligned} \tag{5.73}$$

Using the same approach as in (5.68), we find that

$$\begin{aligned} \overline{v_1 v_1^*} &= \overline{|v_m|^2} + \overline{|v_n|^2} - 2Re\{\overline{v_m v_n^*}\} \\ \overline{v_2 v_2^*} &= \overline{|v_p|^2} + \overline{|v_q|^2} - 2Re\{\overline{v_p v_q^*}\} \\ \overline{v_1 v_2^*} &= \overline{v_m v_p^*} + \overline{v_n v_q^*} - \overline{v_n v_p^*} - \overline{v_m v_q^*} \end{aligned} \tag{5.74}$$

This results in a 2×2 voltage noise correlation matrix representing the desired input and output ports as a function of the elements of $\mathbf{C}_v$. We need not be concerned with the voltages at other terminals, as they are open-circuited. Finally, we convert this to a current noise-correlation matrix,

$$\mathbf{C}_i = \mathbf{Y}\mathbf{C}_v\mathbf{Y}^{*\mathrm{T}} \tag{5.75}$$

where $\mathbf{Y}$ is now the 2×2 reduced admittance matrix representing the two-port, which is generated as part of the linear circuit analysis. Having $\mathbf{C}_i$, we now can find $F_{\min}$, R_n, and $Y_{s,\,\mathrm{opt}}$ by the methods described in Section 5.2.4.

5.3.2.6 Multiport Noise Figure or Temperature

Noise temperature and noise figure can be defined for a multiport, as long as a particular pair of ports has been designated as the input and output. Then, G_t is, of course, the transducer gain between those ports. A greater problem in the multiport case is the treatment of the terminations on the ports that have not been designated input or output. These terminations generate noise, which inevitably finds its way to the output, thus affecting the component's noise temperature. However, these are extrinsic sources, not really part of the component, so in some cases it may be valid to neglect their noise. The correct treatment depends on the particular circumstances, but the former case usually is the correct one.

In the development of Sections 5.2 and 5.3, we have assumed that all terminations are treated as part of the admittance matrix, and the admittance matrix must be formulated as such. If their noise is to be included in the correlation matrices, it should be added in the manner described by (5.56) or (5.57); if not, the noise is simply neglected in generating the correlation matrix. In all cases, the noise of the source and load terminations of the input and output ports must not be included.

References

[5.1] H. Rothe and W. Dahlke, "Theory of Noisy Fourpoles," *Proc. IRE*, Vol. 44, p. 811, 1956.

[5.2] H. Hildebrand and P. Russer, "An Efficient Method for Computer-Aided Noise Analysis of Linear Amplifier Networks," *IEEE Trans. Circuits and Systems*, Vol. CAS-23, p. 235, 1976.

[5.3] V. Rizzoli and A. Lipparini, "Computer Aided Noise Analysis of Linear Multiport Networks of Arbitrary Topology," *IEEE Trans. Microwave Theory Tech.*, Vol. MTT-33, p. 1507, 1985.

[5.4] R. Q. Twiss, "Nyquist's and Thevenin's Theorems Generalized for Nonreciprocal Linear Networks," *Journal of Applied Physics*, Vol. 26, p. 599, 1955.

Chapter 6

Noise Theory of Nonlinear Circuits

One of the great successes of recent years is the development of a workable theory of noise in nonlinear circuits, perhaps inappropriately called *nonlinear noise theory*. The goal of this theory is primarily to describe noise in frequency-conversion circuits, such as mixers and frequency multipliers, and phase noise in oscillators. The use of phase and phase-amplitude modulated digital communication systems has made the latter, in particular, increasingly important.

In this chapter we begin with a brief treatment of nonlinear circuit analysis, as it is necessary for an understanding of nonlinear noise analysis. We then treat modulated noise sources, and finally methods for analyzing noisy nonlinear circuits.

6.1 NONLINEAR CIRCUIT ANALYSIS

Before exploring the subject of nonlinear noise, it is necessary to examine some basics of nonlinear circuit analysis. We must examine methods for performing large-signal analyses of nonlinear circuits, as we must know the large-signal waveforms that affect noise sources that depend upon them. We must also develop methods for treating the noise, which is a small perturbation of the large signals. We examine two methods for large-signal analysis, harmonic-balance analysis and time-domain (or *transient*) analysis.

Central to the treatment of nonlinear noise is the concept of the *conversion matrix*. Conversion matrices describe the response of a circuit, pumped by a large excitation, to small voltage or current deviations. This is precisely the situation when small-signal noise exists in a mixer, frequency multiplier, or oscillator. The idea of a conversion matrix has existed for many years, having been especially useful in the theory of mixers. The purpose of this section is to provide only the necessary basics; a more complete treatment of these topics can be found elsewhere in the literature [6.1].

6.1.1 Harmonic-Balance Analysis

Harmonic-balance analysis is the dominant method for large-signal analysis of microwave circuits. Harmonic-balance analysis is a hybrid method, using elements of both time-domain and frequency-domain analysis. Pure time-domain methods, which predate harmonic balance, do not easily handle circuit elements that are described in the frequency domain. Unfortunately, many linear elements in microwave circuits must be described in the frequency domain, and harmonic balance handles these in a completely natural manner.

In harmonic-balance analysis, the circuit is partitioned into two subcircuits, a linear subcircuit, which is normally represented by a frequency-domain admittance matrix, and a nonlinear subcircuit, which may include both reactive and resistive nonlinearities. The nonlinear subcircuit is analyzed in the time domain. Usually, it is treated as *quasistatic*; that is, the current, charge, or (occasionally) magnetic flux is an instantaneous function of the element's control voltage or current. The quasistatic assumption is not strictly necessary but, at this writing, exists in all harmonic-balance simulators. The quasistatic assumption allows the nonlinear subcircuit to be analyzed algebraically. If the quasistatic requirement were eliminated, the nonlinear subcircuit would have to be described by a set of nonlinear differential equations and integrated in the time domain.

Figure 6.1 shows the partitioned circuit. The linear and nonlinear subcircuits are connected by a number of ports. The port voltages are the independent variables of the system and must generate currents that satisfy Kirchoff's current law at the ports. The voltages can be viewed as either frequency-domain or time-domain quantities. In the frequency domain, we have, for the linear subcircuit,

$$
\begin{bmatrix} \mathbf{I}_{L,1} \\ \mathbf{I}_{L,2} \\ \mathbf{I}_{L,3} \\ \cdots \\ \mathbf{I}_{L,N} \\ \mathbf{I}_{L,N+1} \\ \cdots \end{bmatrix} = \begin{bmatrix} \mathbf{Y}_{1,1} & \mathbf{Y}_{1,2} & \cdots & \mathbf{Y}_{1,N} & \mathbf{Y}_{1,N+1} & \cdots \\ \mathbf{Y}_{2,1} & \mathbf{Y}_{2,2} & \cdots & \mathbf{Y}_{2,N} & \mathbf{Y}_{2,N+1} & \cdots \\ \mathbf{Y}_{3,1} & \mathbf{Y}_{3,2} & \cdots & \mathbf{Y}_{3,N} & \mathbf{Y}_{3,N+1} & \cdots \\ \cdots & \cdots & \cdots & \cdots & \cdots & \cdots \\ \mathbf{Y}_{N,1} & \mathbf{Y}_{N,2} & \cdots & \mathbf{Y}_{N,N} & \mathbf{Y}_{N,N+1} & \cdots \\ \mathbf{Y}_{N+1,1} & \mathbf{Y}_{N+1,2} & \cdots & \mathbf{Y}_{N+1,N} & \mathbf{Y}_{N+1,N+1} & \cdots \\ \cdots & \cdots & \cdots & \cdots & \cdots & \cdots \end{bmatrix} \begin{bmatrix} \mathbf{V}_1 \\ \mathbf{V}_2 \\ \cdots \\ \cdots \\ \mathbf{V}_N \\ \mathbf{V}_{N+1} \\ \cdots \end{bmatrix} \tag{6.1}
$$

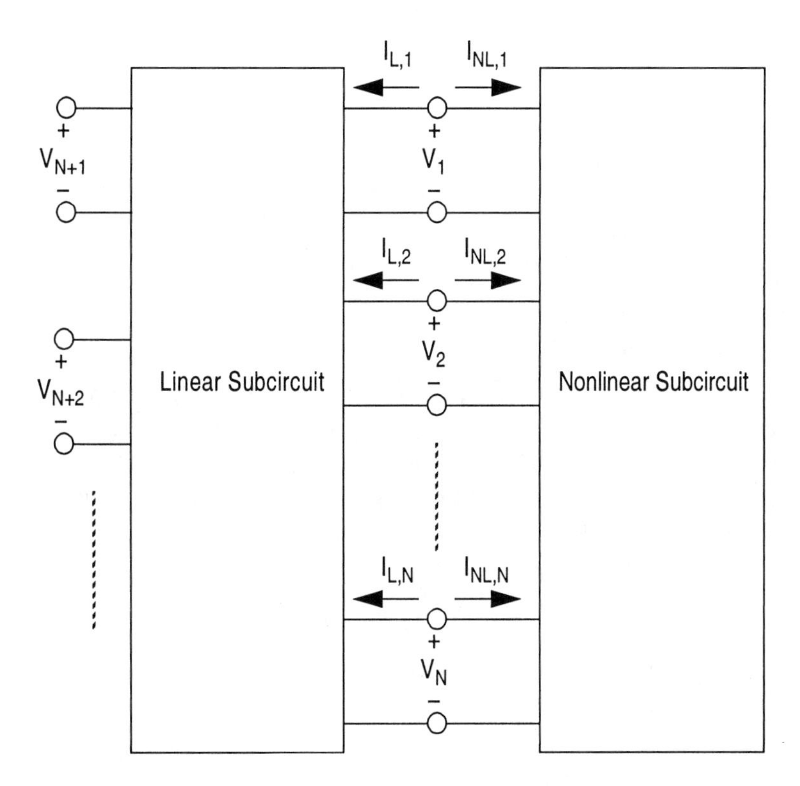

Figure 6.1 In harmonic-balance analysis, the circuit is partitioned into a linear and a nonlinear subcircuit. The voltages at the interface are the state variables of the system; these must satisfy Kirchoff's current law at the interface.

where $\mathbf{V}_n$ and $\mathbf{I}_{L,\,n}$ are vectors of current and voltage at port n and at harmonics of the excitation frequency, $k\omega_p$, from dc to the Kth:

$$\mathbf{I}_{L,\,n} = \begin{bmatrix} I_{n,\,0} \\ I_{n,\,1} \\ \cdots \\ I_{n,\,K} \end{bmatrix} \qquad \mathbf{V}_n = \begin{bmatrix} V_{n,\,0} \\ V_{n,\,1} \\ \cdots \\ V_{n,\,K} \end{bmatrix} \tag{6.2}$$

The terms $\mathbf{Y}_{m,\,n}$ in (6.1) are diagonal submatrices:

$$\mathbf{Y}_{m,\,n} = \mathrm{diag}[Y_{m,\,n}(k\omega_p)], \qquad k = 0, 1, 2, \ldots, K \tag{6.3}$$

Port indices greater than N are excitation ports. The vector of currents in the interface ports, $\mathbf{I}_L$, is

$$\mathbf{I}_L = \hat{\mathbf{Y}}_{1,\,1}\mathbf{V} + \mathbf{Y}_{1,\,2}\mathbf{V}_s = \hat{\mathbf{Y}}_{1,\,1}\mathbf{V} + \mathbf{I}_s \tag{6.4}$$

where $\hat{\mathbf{Y}}_{1,\,1}$ is the submatrix of $\mathbf{Y}$ consisting of the first N rows and columns, and $\hat{\mathbf{Y}}_{1,\,2}$ consists of the same rows and the remaining columns. $\mathbf{V}$ is the vector of voltages at the interface ports and $\mathbf{V}_s$ is the vector of excitations. $\mathbf{I}_s$ is a vector of excitations, transformed to the N interface ports.

At the nonlinear ports, each voltage vector is inverse Fourier transformed to form a time-domain current waveform. The nonlinear-subcircuit current at each port, $I_{\mathrm{NL},\,n}$, is, in general, a function of the voltages at one or more of the interface ports:

$$I_{\mathrm{NL},\,n}(t) = f_n(V_l(t), V_m(t), \ldots) \tag{6.5}$$

where l, m, n are port indices. The frequency domain current at the port, $\mathbf{I}_{\mathrm{NL},\,n}$, is found simply by Fourier transformation:

$$\mathcal{F}\{I_{\mathrm{NL},\,n}(t)\} \rightarrow \mathbf{I}_{\mathrm{NL},\,n} \tag{6.6}$$

Kirchoff's current law must be satisfied. From (6.4) and (6.6), we have

$$\mathbf{F(V)} = \hat{\mathbf{Y}}_{1,1}\mathbf{V} + \mathbf{I}_s + \mathbf{I}_{\mathrm{NL}} = \mathbf{0} \tag{6.7}$$

$\mathbf{F(V)}$ is called the *current-error vector.* $\mathbf{F(V)} = \mathbf{0}$ when the correct set of port voltages has been determined; however, during the iterative solution process it has a nonzero value. The solution process must find a set of voltages $\mathbf{V}$ that satisfies (6.7).

Equation (6.7) tells us that we must find, simultaneously, the zeros of a large number of functions of an equally large number of variables. That is, we have $N(2K + 1)$ real and imaginary voltage components and the same number of equations. Various methods have been developed for finding the zeros of a set of multidimensional functions. After much research throughout the past three decades, a consensus has emerged that the preferred method is multidimensional Newton iteration. This involves first estimating the zero of $\mathbf{F(V)}$, then improving the estimate iteratively until the result is adequately close to the solution. At each iteration we must find $\Delta\mathbf{V}$ by solving

$$\mathbf{F(V)} - \frac{d}{d\mathbf{V}}\mathbf{F(V)}\bigg|_{\mathbf{V} = \mathbf{V}_p} \Delta\mathbf{V} = \mathbf{0} \tag{6.8}$$

where $\mathbf{V}_p$ is the port-voltage vector at some particular iteration. We then update $\mathbf{V}$ for the $p + 1$th iteration:

$$\mathbf{V}_{p+1} = \mathbf{V}_p - \Delta\mathbf{V} \tag{6.9}$$

The term $d\mathbf{F(V)}/d\mathbf{V}$ is a Jacobian matrix. It is closely related to a conversion matrix, which we examine in Section 6.1.3. The Jacobian contains derivatives of each current component with respect to each harmonic component, so it inherently provides information about how the current error, at each harmonic, is affected by each voltage harmonic. In the case of resistive or charge nonlinearities, the terms are simply Fourier components of the conduction or charge waveforms. Each iteration involves solving a matrix equation; the size of the matrix is equal to the number of voltage variables, $N(2K + 1)$. In large circuits, this becomes a huge computational task.

Harmonic-balance analysis is conceptually relatively simple. Making it practical is more difficult. A practical harmonic-balance solver must be reasonably efficient and have strong convergence properties. It must also accommodate multitone excitations, oscillatory circuits, and other types of problems somewhat removed from the simple, single-tone method described above. See the references for more information on these matters.

6.1.2 Time-Domain Analysis

Another method for performing large-signal analysis is integration in the time domain [6.2]. Time-domain analysis predates harmonic balance analysis and remains the dominant method for RF circuits, operating at frequencies where distributed effects are of less concern.

Any circuit having N independent reactive elements can be described by an nth order differential equation. That differential equation can also be expressed in matrix form as N first-order equations. It becomes, in general,

$$\begin{bmatrix} C_N & & & \\ & C_{N-1} & & \\ & & \cdots & \\ & & & C_1 \end{bmatrix} \begin{bmatrix} \dot{X}_N \\ \dot{X}_{N-1} \\ \cdots \\ \dot{X}_1 \end{bmatrix} + \begin{bmatrix} f_N(X_1, X_2, \ldots, X_N) \\ f_{N-1}(X_1, X_2, \ldots, X_N) \\ \cdots \\ f_1(X_1, X_2, \ldots, X_N) \end{bmatrix} + \begin{bmatrix} U_N \\ U_{N-1} \\ \cdots \\ U_1 \end{bmatrix} = \begin{bmatrix} 0 \\ 0 \\ 0 \\ 0 \end{bmatrix} \tag{6.10}$$

where C_n are constants, $f_n(\ldots)$ are, in general, nonlinear functions, U_n are excitations (most of which are zero), X_n are either voltage or current functions of time, and the dot over X_n indicates time differentiation. Equation (6.10) can be expressed as

$$\mathbf{C}\dot{\mathbf{X}} + \mathbf{F}(\mathbf{X}) + \mathbf{U} = \mathbf{0} \tag{6.11}$$

where boldface characters, as before, indicate matrices or vectors.

Equation (6.11) can be integrated in a number of ways. One simple approach is to use a forward-difference formula,

$$\dot{\mathbf{X}} = (1/\Delta t)(\mathbf{X}_{t+\Delta t} - \mathbf{X}_t) = -\mathbf{C}^{-1}(\mathbf{F}(\mathbf{X}_{t+\Delta t}) + \mathbf{U}) \tag{6.12}$$

which is solved for $\mathbf{X}_{t+\Delta t}$, the vector of unknowns at the next time point. The quantity $\mathbf{X}_t$ is known; it is the solution at the previous time point or, at the beginning of the integration, the initial condition. At each time step, an iterative procedure, usually multidimensional Newton iteration, solves

(6.12). The process continues, point to point, through the required time period.

Time-domain integration requires that all quantities be expressed in the time domain. In microwave circuits, many circuit elements are best described in the frequency domain, and dealing with such elements is a significant complication. One possibility is to convert frequency domain quantities to the time domain by Fourier transformation, to obtain an impulse response. Another is to convert the frequency-domain data into an expression in the Laplace domain, which can be inverted into the time domain by one of several methods.

Another difficulty in time-domain analysis is the need to obtain the sinusoidal, steady-state response of the circuit. Unlike harmonic-balance analysis, which calculates the steady-state response directly, time-domain analysis calculates the transient response. To obtain the steady-state response, one must either integrate through the transient response, which may be very long, or use techniques such as *shooting methods* to search for the steady state directly. None of these problems are insuperable, but they do complicate the solution process, create the possibility of nonconvergence to a solution, and increase computational cost.

6.1.3 Conversion Matrices

Frequently, we encounter problems in which a mixture of large and small signals are applied to a nonlinear circuit. In a mixer, for example, the large signal is the local oscillator, and the small signals are the RF excitation and noise. The small-signal response can be treated as small, *linear*, perturbations of the large-signal response, creating current and voltage sidebands around the large-signal harmonics. We wish to develop an expression that relates the voltages and currents at those sidebands. The method for doing this is called *conversion matrix analysis.*

We consider a nonlinear, resistive, two-terminal circuit element having the I/V relationship $I = f(V)$. V consists of a large-signal, time-varying component, $V_L(t)$, and a small-signal component, $v(t)$. We assume that the large-signal waveform is known; it can be calculated by harmonic-balance or time-domain methods. Our main concern is to determine the noise, which clearly is a small perturbation of the large-signal response.

The current resulting from the small-signal excitation is found by expanding $f(V_L + v)$ in a Taylor series and subtracting the large-signal component. The Taylor series is

$$f(V_L + v) = f(V_L) + \frac{d}{dV} f(V)\bigg|_{V = V_L(t)} v + \frac{1}{2} \frac{d^2}{dV^2} f(V)\bigg|_{V = V_L(t)} v^2 + \frac{1}{6} \frac{d^3}{dV^3} f(V)\bigg|_{V = V_L(t)} v^3 + \ldots \tag{6.13}$$

The small-signal, incremental current is

$$i(v) = I(V_L + v) - I(V_L) = f(V_L + v) - f(V_L) \tag{6.14}$$

Since $v \ll V_L$, the nonlinear terms, v^2, v^3, ... are negligible; then,

$$i(t) = \frac{d}{dV} f(V)\bigg|_{V = V_L(t)} v(t) \tag{6.15}$$

which can be expressed as

$$i(t) = g(t)v(t) \tag{6.16}$$

where

$$g(t) = \frac{d}{dV} f(V)\bigg|_{V = V_L(t)} \tag{6.17}$$

We have effectively linearized the I/V function about a time-varying central value, $V_L(t)$. Similarly, for a current-controlled resistor with the V/I characteristic

$$V = f_R(I) \tag{6.18}$$

the small-signal v/i relation is

$$v(t) = r(t)i(t) \tag{6.19}$$

where

$$r(t) = \frac{d}{dI} f_R(I) \Big|_{I = I_L(t)} \tag{6.20}$$

Often, the nonlinear element's current is a function of two voltages, $I = f_2(V_1, V_2)$. This function can be expanded in a two-dimensional Taylor series. We obtain, simply,

$$i(t) = g_1(t)v_1(t) + g_2(t)v_2(t) \tag{6.21}$$

where

$$g_1(t) = \frac{\partial}{\partial V_1} f_2(V_1, V_2) \Bigg|_{\substack{V_1 = V_{L,1}(t) \\ V_2 = V_{L,2}(t)}} \tag{6.22}$$

and

$$g_2(t) = \frac{\partial}{\partial V_2} f_2(V_1, V_2) \Bigg|_{\substack{V_1 = V_{L,1}(t) \\ V_2 = V_{L,2}(t)}} \tag{6.23}$$

showing that our incremental, doubly controlled nonlinear conductance is equivalent to two elements, at least one of which is a transconductance. When the I / V characteristic is a function of more than two voltages, we find, predictably, that

$$i(t) = g_1(t)v_1(t) + g_2(t)v_2(t) + g_3(t)v_3(t) + \ldots \tag{6.24}$$

Some, or even all, of the g_n can be transconductances.

The treatment of capacitors is similar. The small-signal charge function of a nonlinear capacitor having the Q / V characteristic $Q = f_Q(V)$ is

$$q(t) = \frac{d}{dV} f_Q(V) \Big|_{V = V_L(t)} v(t) \tag{6.25}$$

or

$$q(t) = c(t)v(t) \tag{6.26}$$

The capacitor's current is the time derivative of the charge:

$$i(t) = \frac{d}{dt}q(t) = c(t)\frac{d}{dt}v(t) + v(t)\frac{d}{dt}c(t) \tag{6.27}$$

Charge in a nonlinear capacitor can also be a function of multiple voltages. Then, the small-signal charge is

$$q(t) = c_1(t)v_1(t) + c_2(t)v_2(t) + c_3(t)v_3(t) + \ldots \tag{6.28}$$

The small-signal current is the time derivative of the small-signal charge:

$$\begin{aligned} i(t) = \frac{d}{dt}q(t) &= c_1(t)\frac{d}{dt}v_1(t) + v_1(t)\frac{d}{dt}c_1(t) \\ &+ c_2(t)\frac{d}{dt}v_2(t) + v_2(t)\frac{d}{dt}c_2(t) + \ldots \end{aligned} \tag{6.29}$$

Because this is a linear perturbation, the small-signal quantity does not generate harmonics, but the large-signal may. The small-signal mixing frequencies are therefore $\omega = \pm\omega_1 + k\,\omega_p$, where ω_1 is the small-signal excitation frequency, ω_p is the large-signal fundamental frequency, and k are integers. The set of frequency components is shown in Figure 6.2(a). The frequencies consist of two tones on either side of each large-signal harmonic, separated by $\omega_0 = |\omega_1 - \omega_p|$. The mixing frequencies can also be expressed as

$$\omega_k = \omega_0 + k\omega_p \tag{6.30}$$

which includes only the negative-frequency components of the lower sidebands and the positive-frequency components of the upper sidebands. This set of frequencies, shown in Figure 6.2(b), has all the information of the set in Figure 6.2(a); negative- and positive-frequency components are complex conjugate pairs, so knowledge of only one is necessary.

The conversion matrix is nothing more than (6.16), (6.19), or (6.26) expressed in the frequency domain. For a time-varying conductance, we have

$$v'(t) = \sum_{k=-\infty}^{\infty} V_k \exp(j\omega_k t) \tag{6.31}$$

and

$$i'(t) = \sum_{k=-\infty}^{\infty} I_k \exp(j\omega_k t) \tag{6.32}$$

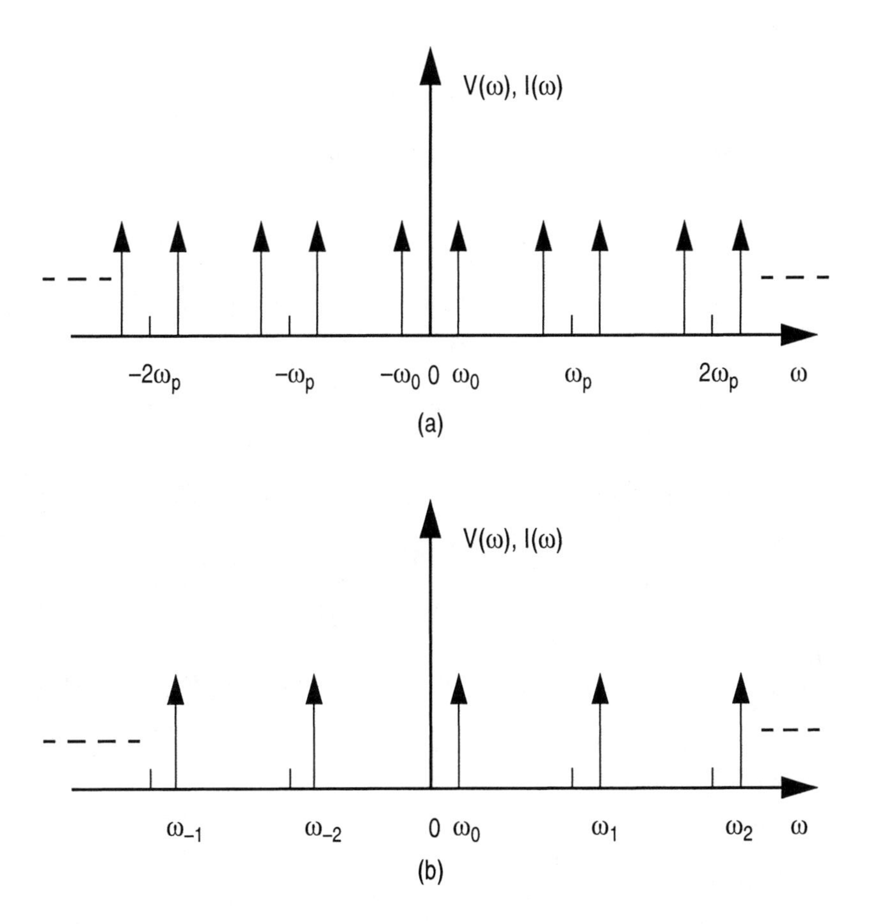

Figure 6.2 (a) Spectrum of small-signal mixing frequencies in the pumped nonlinear element; (b) spectrum of frequency components given by (6.30).

The primes indicate that $v'(t)$ and $i'(t)$ are sums of positive- and negative-frequency phasor components in (6.30), not the complete time waveforms. The conductance waveform $g(t)$ is expressed by its Fourier series,

$$g(t) = \sum_{k=-\infty}^{\infty} G_k \exp(jk\omega_p t) \tag{6.33}$$

Substituting (6.31) through (6.33) into (6.16) gives the relation,

$$\sum_{k=-\infty}^{\infty} I_k \exp(j\omega_k t) = \sum_{n=-\infty}^{\infty} \sum_{m=-\infty}^{\infty} G_n V_m \exp(j\omega_{m+n} t) \tag{6.34}$$

Equating terms on both sides of (6.34) and limiting the summations to K results in a set of linear equations:

$$\begin{bmatrix} I^*_{-K} \\ I^*_{-K+1} \\ I^*_{-K+2} \\ \cdots \\ I^*_{-1} \\ I_0 \\ I_1 \\ \cdots \\ I_K \end{bmatrix} = \begin{bmatrix} G_0 & G_{-1} & G_{-2} & \cdots & G_{-2K} \\ G_1 & G_0 & G_{-1} & \cdots & G_{-2K+1} \\ G_2 & G_1 & G_0 & \cdots & G_{-2K+2} \\ \cdots & \cdots & \cdots & \cdots & \cdots \\ G_{K-1} & G_{K-2} & G_{K-3} & \cdots & G_{-K-1} \\ G_K & G_{K-1} & G_{K-2} & \cdots & G_{-K} \\ G_{K+1} & G_K & G_{K-1} & \cdots & G_{-K+1} \\ \cdots & \cdots & \cdots & \cdots & \cdots \\ G_{2K} & G_{2K-1} & G_{2K-2} & \cdots & G_0 \end{bmatrix} \begin{bmatrix} V^*_{-K} \\ V^*_{-K+1} \\ V^*_{-K+2} \\ \cdots \\ V^*_{-1} \\ V_0 \\ V_1 \\ \cdots \\ V_K \end{bmatrix} \tag{6.35}$$

or, in matrix notation,

$$\mathbf{I} = \mathbf{G}\mathbf{V} \tag{6.36}$$

The conjugates of the negative-frequency components are caused by a change of definition; according to (6.30), ω_k is negative when $k < 0$, so the

I_k and V_k are negative-frequency components when $k < 0$. We would rather define them as positive-frequency phasors, so we must introduce the conjugates.

We obtain a similar result for a resistor:

$$\begin{bmatrix} V^*_{-K} \\ V^*_{-K+1} \\ V^*_{-K+2} \\ \cdots \\ V^*_{-1} \\ V_0 \\ V_1 \\ \cdots \\ V_K \end{bmatrix} = \begin{bmatrix} R_0 & R_{-1} & R_{-2} & \cdots & R_{-2K} \\ R_1 & R_0 & R_{-1} & \cdots & R_{-2K+1} \\ R_2 & R_1 & R_0 & \cdots & R_{-2K+2} \\ \cdots & \cdots & \cdots & \cdots & \cdots \\ R_{K-1} & R_{K-2} & R_{K-3} & \cdots & R_{-K-1} \\ R_K & R_{K-1} & R_{K-2} & \cdots & R_{-K} \\ R_{K+1} & R_K & R_{K-1} & \cdots & R_{-K+1} \\ \cdots & \cdots & \cdots & \cdots & \cdots \\ R_{2K} & R_{2K-1} & R_{2K-2} & \cdots & R_0 \end{bmatrix} \begin{bmatrix} I^*_{-K} \\ I^*_{-K+1} \\ I^*_{-K+2} \\ \cdots \\ I^*_{-1} \\ I_0 \\ I_1 \\ \cdots \\ I_K \end{bmatrix} \tag{6.37}$$

R_n are the Fourier components of the resistance waveform. In matrix form, (6.37) is

$$\mathbf{V} = \mathbf{RI} \tag{6.38}$$

For a given conductance/resistance element,

$$\mathbf{R} = \mathbf{G}^{-1} \tag{6.39}$$

Clearly, both **R** and **G** must exist and be nonsingular. The easiest way to violate this condition is for $r(t)$ or $g(t)$ to be zero or infinite for some period in the waveform.

To find the conversion matrix of a capacitor, we express the capacitance waveform $c(t)$ by its Fourier series

$$c(t) = \sum_{k=-\infty}^{\infty} C_k \exp(jk\omega_p t) \tag{6.40}$$

The current is

$$i'(t) = \frac{d}{dt}q'(t) \tag{6.41}$$

The charge $q'(t)$ has the form

$$q'(t) = \sum_{k=-\infty}^{\infty} Q_n \exp(j\omega_k t) \tag{6.42}$$

and the voltage is expressed, as before, by (6.31). Substituting (6.31), (6.40), and (6.42) into (6.26) gives

$$\sum_{k=-\infty}^{\infty} Q_k \exp(j\omega_k t) = \sum_{n=-\infty}^{\infty} \sum_{m=-\infty}^{\infty} C_n V_m \exp(j\omega_{m+n} t) \tag{6.43}$$

The current can be found by differentiating in the frequency domain, which corresponds to multiplying by $j\omega$:

$$\sum_{k=-\infty}^{\infty} I_k \exp(j\omega_k t) = \sum_{n=-\infty}^{\infty} \sum_{m=-\infty}^{\infty} j\omega_{m+n} C_n V_m \exp(j\omega_{m+n} t) \tag{6.44}$$

Equating terms at the same frequency gives the matrix equation

$$\mathbf{I} = j\Omega\mathbf{C}\mathbf{V} \tag{6.45}$$

and $\mathbf{C}$ has a form similar to (6.35) and (6.37). The matrix Ω is diagonal with elements $j\omega_{-K}$ to $j\omega_K$:

$$\Omega = \begin{bmatrix} \omega_{-K} & 0 & \dots & 0 \\ 0 & \omega_{-K+1} & \dots & 0 \\ \dots & \dots & \dots & \dots \\ 0 & 0 & \dots & \omega_K \end{bmatrix} \tag{6.46}$$

For completion, we note that a linear admittance can have a conversion-matrix representation. Such a matrix is, of course, simply a diagonal. In general, it is

$$\mathbf{Y} = \begin{bmatrix} Y(\omega_{-K}) & 0 & \dots & 0 \\ 0 & Y(\omega_{-K+1}) & \dots & 0 \\ \dots & \dots & \dots & \dots \\ 0 & 0 & \dots & Y(\omega_K) \end{bmatrix} \qquad (6.47)$$

where $Y(\omega_k)$ is simply the admittance of the element at the mixing frequency ω_k. Note that $Y(\omega_k)$ must be conjugate for $k < 0$. For a simple conductance G, (6.47) is a diagonal whose elements are G. Similarly, a capacitor of capacitance C can be expressed by (6.45) where $\mathbf{C} = \text{diag}(C)$. Finally, a linear N-port admittance matrix has the form

$$\mathbf{Y} = \begin{bmatrix} \mathbf{Y}_{1,1} & \mathbf{Y}_{1,2} & \cdots & \mathbf{Y}_{1,N} \\ \mathbf{Y}_{2,1} & \mathbf{Y}_{2,2} & \cdots & \mathbf{Y}_{2,N} \\ \cdots & \cdots & \cdots & \cdots \\ \mathbf{Y}_{N,1} & \mathbf{Y}_{N,2} & \cdots & \mathbf{Y}_{N,N} \end{bmatrix} \qquad (6.48)$$

where

$$\mathbf{Y}_{n,m} = \begin{bmatrix} Y_{n,m}(\omega_{-K}) & 0 & \dots & 0 \\ 0 & Y_{n,m}(\omega_{-K+1}) & \dots & 0 \\ \dots & \dots & \dots & \dots \\ 0 & 0 & \dots & Y_{n,m}(\omega_K) \end{bmatrix} \qquad (6.49)$$

Again, when $k < 0$, $Y_{n,m}(\omega_k)$ must be the complex conjugate of the positive-frequency value.

6.1.4 Circuit Analysis with Conversion Matrices

It is a relatively straightforward matter to create a small-signal admittance representation of a pumped nonlinear circuit using conversion matrices. Since the perturbation of the circuit described by the conversion matrix is linear, the admittance matrix can be assembled in a manner entirely analogous to the method described in Section 5.3.1. The difference, of course, is that the elements in the matrix are conversion submatrices, not scalars.

For example, consider the simple circuit of Figure 6.3, which could represent a diode mixer. The time-varying conductance, $g(t)$, has been determined from a large-signal analysis of the pumped diode, according to (6.15) to (6.17), and its conversion matrix $\mathbf{G}$ has been formulated, following (6.31) to (6.36). The nodal admittance matrix is created in the same manner as the scalar one (Section 5.3.1). It can be written by inspection:

$$\mathbf{Y} = \begin{bmatrix} \mathbf{Y}_1 & 0 & -\mathbf{Y}_1 \\ 0 & \mathbf{Y}_2 & -\mathbf{Y}_2 \\ -\mathbf{Y}_1 & -\mathbf{Y}_2 & \mathbf{Y}_1 + \mathbf{Y}_2 + \mathbf{G} \end{bmatrix} \tag{6.50}$$

where $\mathbf{G}$, $\mathbf{Y}_1$, and $\mathbf{Y}_2$ are conversion submatrices of the forms in (6.35) and (6.47). For mixer analysis, we reduce this matrix to a two-port, where port 1 is between node 1 and ground, and port 2 is between node 2 and ground. This is accomplished by inverting the admittance matrix to form an imped-

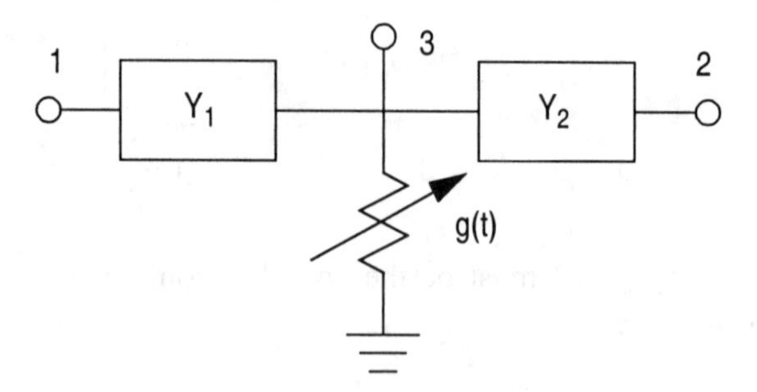

Figure 6.3 Simple circuit representing a diode mixer.

ance matrix, then deleting rows and columns associated with node 3, thus open-circuiting it.[1] The result is

$$\mathbf{Y} = \begin{bmatrix} \mathbf{Y}_{1,1} & \mathbf{Y}_{1,2} \\ \mathbf{Y}_{2,1} & \mathbf{Y}_{2,2} \end{bmatrix} \tag{6.51}$$

This matrix still has the dimension $4N + 2$, where $2N$ is the maximum harmonic used in determining the conversion submatrices. We wish to reduce it to a 2×2 matrix relating a particular port 2 frequency (say, the IF, ω_0) to a particular port 1 frequency (the RF, ω_1). First, however, we should add the out-of-band port terminations at mixing frequencies other than ω_1 at port 1 and ω_0 at port 2.[2] Those terminations, $Y_{t1}(\omega_k)$ at port 1 and $Y_{t2}(\omega_k)$ at port 2, can be represented by matrices of the form (6.47). Because we do not want to terminate the ports at the input and output frequencies, $Y_{t1}(\omega_1) = Y_{t2}(\omega_0) = 0$. The admittance matrix becomes

$$\mathbf{Y} = \begin{bmatrix} \mathbf{Y}_{1,1} + \mathbf{Y}_{t1} & \mathbf{Y}_{1,2} \\ \mathbf{Y}_{2,1} & \mathbf{Y}_{2,2} + \mathbf{Y}_{t2} \end{bmatrix} \tag{6.52}$$

where $\mathbf{Y}_{t1}$ and $\mathbf{Y}_{t2}$ are the termination matrices, in the form of (6.47). Finally, to eliminate the terminated ports, we invert the terminated admittance matrix, forming an impedance matrix, and set all rows and columns to zero except the ones corresponding to ω_1 at port 1 and ω_0 at port 2. The result is a 2×2 impedance matrix, which can at last be converted to any other type of 2×2 matrix, such as an admittance matrix or a scattering matrix:

$$\begin{bmatrix} I_1(\omega_1) \\ I_2(\omega_0) \end{bmatrix} = \begin{bmatrix} Y_{1,1} & Y_{1,2} \\ Y_{2,1} & Y_{2,2} \end{bmatrix} \begin{bmatrix} V_1(\omega_1) \\ V_2(\omega_0) \end{bmatrix} \tag{6.53}$$

This matrix is a linear admittance matrix. For mathematical purposes, the fact that the quantities are at different frequencies is quite irrelevant.

1. A full matrix inversion, while intuitively useful for explaining this process, is not numerically efficient. See Section 5.3.1.2 for a better method.

2. The handling of such terminations is occasionally confusing and equally controversial. We examine this matter further in Chapter 9.

This matrix can be interpreted and manipulated in the same manner as any other two-port admittance matrix, providing port impedances, conversion loss, and similar quantities.

6.2 NOISE SOURCE CHARACTERIZATION

In nonlinear circuits, we have two fundamental concerns. The first is to determine the effect of noise in *nonautonomous* circuits, circuits such as mixers and frequency multipliers, which have external RF excitations. Examples are the noise figure of a mixer or phase noise in a frequency multiplier. The second concern is noise in oscillatory circuits having no external excitation, other than dc bias. Such circuits are called *autonomous*.

Solid-state devices may have static noise sources, such as thermal noise from parasitic resistances. Other noise sources, however, are functions of the large-signal currents or voltages in the device, and thus are modulated by large-signal voltage or current waveforms. The modulation process, as one might expect, creates noise sidebands around each large-signal harmonic. Less intuitive is the fact that it introduces correlations between those sideband components. Accounting for those correlations is critical to accurate noise analysis of large-signal circuits.

6.2.1 Noise Sources in Nonlinear Circuits

In the linear noise analysis of Chapter 5, we considered only the response of a linear circuit to one or more noise sources. Because the circuit was linear, we were able to ignore sinusoidal or other deterministic excitations. If we were interested in them, we needed only to find the circuit response to those excitations and add them to the noise.

In a nonlinear circuit, the situation is not so simple. We are faced with a circuit that is driven by one or more large signals, and small-signal noise is present as well. Some of the noise sources are conventional ones, such as the thermal noise sources associated with resistors, which we shall call *static* or *unmodulated* noise sources. Many, however, are more complex; frequently, their noise levels depend on large-signal control voltages or currents somewhere in the circuit. For example, Section 4.2.1 shows that the spectral density of the shot noise current in a Schottky-barrier diode, $S_i(f)$, is

$$S_i(f) = 2qI_j \tag{6.54}$$

where q is the electron charge and I_j is the junction current. In linear circuits, I_j is simply a dc quantity, the bias current in the device, but in nonlinear circuits $I_j = I_j(t)$ is generally a large-signal waveform. $I_j(t)$ can be determined by either harmonic-balance analysis, most commonly used for microwave circuits, or time-domain methods. Clearly, $S_i(f)$ is modulated by the large-signal junction current. Such pumped or modulated noise sources have been called *cyclostationary*; while their statistics are, strictly, nonstationary, they remain the same from cycle to cycle.

We furthermore assume that all the modulated sources treated in this chapter are *quasistatic*; that is, the appropriate statistics of the noise (i.e., average, mean-square value, and spectral density) change instantaneously with changes in the control current or voltage. This assumption is not unlike the quasistatic assumption used in modeling nonlinear circuit elements (see, for example, [6.1]) and is subject to the same limitations. Nonquasistatic behavior of solid-state devices is well known and documented; however, there has not been enough research in modulated noise sources, at this writing, to identify the limits to this assumption when applied to noise analysis.

6.2.2 Modulated Noise Sources

It is well known that narrowband samples of a broadband noise source, at different frequencies, are uncorrelated. That is, if we were to filter the output of a broadband microwave noise source into two narrow channels centered at, say, 1 GHz and 2 GHz, those two noise processes would be uncorrelated. This situation may be surprising, at first, since the same source was used to generate both processes. It is easily proven by expressing the noise processes $n_1(t)$ and $n_2(t)$, which can represent either voltages or currents, as

$$\begin{aligned} n_1(t) &= a_1(t)\cos(\omega_1 t) - b_1(t)\sin(\omega_1 t) \\ n_2(t) &= a_2(t)\cos(\omega_2 t) - b_2(t)\sin(\omega_2 t) \end{aligned} \tag{6.55}$$

where $a_{1,2}(t)$ and $b_{1,2}(t)$ are baseband random processes (Section 2.2.1) and ω_1, ω_2 are reference frequencies, usually the center frequencies of the noise spectra. Forming $\overline{n_1(t)n_2(t)}$ shows immediately that the correlation is zero.

When a noise source is modulated by a large-signal function, the situation is not so simple. To examine such sources, we begin by assuming that a noise source has a spectral density given by

$$S(f) = F_n(f, I) \tag{6.56}$$

where F_n describes the functional dependence of the noise spectrum on frequency and I, the control quantity. We risk a certain lack of generality by our use of the current symbol I, to represent the control quantity. However, virtually all of the noise sources we encounter in practical microwave circuits are current controlled, and, if we were to encounter a voltage-controlled one, the extension should be obvious. Similarly, the treatment of a noise source having multiple control currents (or voltages) is a minor extension of the following derivation.

In a modulated source, $I = I(t)$, a large-signal function and, as noted earlier, we assume F_n to be quasistatic in I. Then

$$S(f, t) = F_n(f, I(t)) \tag{6.57}$$

This can be converted into the form

$$S(f, t) = f_m^2(t) F_n(f, I_0) \tag{6.58}$$

where the modulation function, $f_m(t)$, is

$$f_m(t) = \sqrt{\frac{F_n(f, I(t))}{F_n(f, I_0)}} \tag{6.59}$$

and I_0 is some convenient value of $I(t)$, often its time average. We have implicitly assumed in (6.59) that the shape of the noise spectrum is not modified by the modulation. This assumption is not particularly restrictive, as we inevitably deal with incremental bandwidths; if the assumption is not valid, we simply use a different $f_m(t)$ function at different frequencies. Or, in other words, we could have written $f_m(t)$ as $f_m(f, t)$, where f is the frequency at which (6.59) was evaluated.

The modulated noise process itself, $n_m(t)$, is

$$n_m(t) = f_m(t) n_{I_0}(t) \tag{6.60}$$

where $n_{I_0}(t)$ is the noise process having the spectrum defined by $F_n(f, I_0)$.

Example: Shot Noise

Consider shot noise in a mixer diode. The noise spectral density is given by (6.54), and the diode's junction current, $I_j(t)$, is a train of pulses. From (6.56) to (6.59), we can express the spectrum as

$$S(f, t) = f_m^2(t) \cdot 2qI_0 \tag{6.61}$$

where

$$f_m(t) = \sqrt{\frac{I_j(t)}{I_0}} \tag{6.62}$$

In this case, the use of the normalizing quantity I_0 seems unnecessary; we could easily have formulated

$$\begin{aligned} S(f, t) &= f_m^2(t) \cdot 2q \\ f_m(t) &= \sqrt{I_j(t)} \end{aligned} \tag{6.63}$$

The form of (6.58) to (6.59) is sometimes necessary, however, to allow more complex forms of the modulated noise source. In some cases it may also improve the numerical conditioning of the resulting matrices.

Example: Time-Varying Resistor

From Section 2.3.1, a resistor's noise spectrum is

$$S(f) = 4KT\Delta fG \tag{6.64}$$

Some noise sources, such as the channel of a FET used as a resistive mixer, are best modeled as time-varying conductances. Then

$$S(f, t) = 4KT\Delta fG(t) \tag{6.65}$$

Expressing this in the form of (6.58) gives

$$S(f,t) = f_m^2(t) \cdot 4KT\Delta f G_0$$
$$f_m(t) = \sqrt{\frac{G(t)}{G_0}} \tag{6.66}$$

where G_0 is some convenient value of $G(t)$, often $\overline{G(t)}$.

6.2.3 Correlation Matrix of a Modulated Noise Source

Equation (6.58) describes a simple modulation process, in which the noise spectrum given by F_n is modulated by the function $f_m(t)$. As with any modulation process, the result is a number of harmonics of the large-signal quantity with sidebands mirrored above and below each harmonic. Less obvious is the fact that, in contrast to the linear case, those sidebands are correlated. Thus, we need to find a frequency-domain correlation matrix that describes the correlations between those quantities.

In Section 6.2.2 we showed that a modulated noise source can be described as a modulating function that multiplies a simple unmodulated noise process. For a current-noise source,

$$i_m(t) = f_m(t)\, i_n(t) \tag{6.67}$$

where $i_m(t)$ is the modulated noise process, $i_n(t)$ is an unmodulated one, and $f_m(t)$ is the modulation waveform. These are described in (6.56) to (6.59). Because of the obvious similarity between (6.16) and (6.67), we can represent (6.67) in the frequency domain by a conversion matrix:

$$\mathbf{I}_m = \mathbf{F}_m \mathbf{I}_n \tag{6.68}$$

The modulation function is described by a Fourier series, truncated to K harmonics:

$$f_m(t) = \sum_{k=-K}^{K} F_k \exp(jk\omega_p t) \tag{6.69}$$

The conversion-matrix representation of $f_m(t)$ becomes

$$\mathbf{F}_m = \begin{bmatrix} F_0 & F_{-1} & F_{-2} & \dots & F_{-2K} \\ F_1 & F_0 & F_{-1} & \dots & F_{-2K+1} \\ F_2 & F_1 & F_0 & \dots & F_{-2K+2} \\ \dots & \dots & \dots & \dots & \dots \\ F_{2K} & F_{2K-1} & F_{2K-2} & \dots & F_0 \end{bmatrix} \tag{6.70}$$

and $\mathbf{I}_n$, $\mathbf{I}_m$, as before, have the form

$$\mathbf{I} = \begin{bmatrix} i^*_{-K} \\ i^*_{-K+1} \\ i^*_{-K+2} \\ \dots \\ i^*_{-1} \\ i_0 \\ i_1 \\ \dots \\ i_K \end{bmatrix} \tag{6.71}$$

As with static sources, the current noise correlation matrix, $\mathbf{C}_i$, is

$$\mathbf{C}_i = \overline{\mathbf{I}_m \mathbf{I}_m^{*\mathrm{T}}} \tag{6.72}$$

Substituting (6.68) into (6.72) gives

$$\mathbf{C}_i = \overline{(\mathbf{F}_m \mathbf{I}_n) \cdot (\mathbf{F}_m \mathbf{I}_n)^{*\mathrm{T}}} = \mathbf{F}_m \overline{\mathbf{I}_n \mathbf{I}_n^{*\mathrm{T}}} \mathbf{F}_m^{*\mathrm{T}} \tag{6.73}$$

The noise source $i_n(t)$ is unmodulated, so $\mathbf{I}_n$ is a diagonal matrix,

$$\overline{\mathbf{I}_n \mathbf{I}_n^{*\mathrm{T}}} = \mathrm{diag}(\overline{|i_k|^2}) \tag{6.74}$$

and its terms are real. It should be clear from (6.73) that the correlation matrix $\mathbf{C}_i$ is dense, indicating that frequency components at various sidebands are correlated. These correlation properties must be included in the noise analysis of pumped nonlinear circuits.

Of course, not all noise sources in a nonlinear circuit are modulated. The correlation matrix of an unmodulated noise source, in the form of (6.72) to (6.74), is easily determined. In that case $f_m(t) = 1$ so $\mathbf{F}_m$ is simply the identity matrix. Then

$$\mathbf{C}_i = \overline{\mathbf{I}_n \mathbf{I}_n^{*\mathrm{T}}} \tag{6.75}$$

$\mathbf{C}_i$ is a diagonal matrix whose terms are the noise source's mean square values at the frequencies ω_k.

6.3 NOISE IN NONAUTONOMOUS CIRCUITS

The analysis of noise in nonautonomous circuits is not unlike the calculation of conversion loss in a mixer. A conversion matrix is used to linearize the circuit around its large-signal operating conditions, and the small-signal noise is treated as an excitation. We must find the correlation matrix of the noise at the output port; the mean-square noise output voltage or current is simply one component of that matrix. With this information, the noise output power is easily calculated and converted, through the relations in Chapter 3, to a noise temperature or noise figure.

6.3.1 Noise Source Correlation Matrix

Figure 6.4 shows the problem we address. The figure shows a circuit, described by a nodal admittance matrix, and a number of modulated noise sources connected to some, but not necessarily all, of the nodes. We could include among these sources simple, unmodulated noise sources, such as those associated with an ordinary resistor. From (6.74), the correlation matrix of such a source is simply a diagonal. (We address the problem of including correlated linear sources momentarily.)

The correlation matrix of each source, $\mathbf{C}_{ip}$, is given by (6.72) to (6.74), where p is the port number. The complete correlation matrix of the full set of sources in Figure 6.4, $\mathbf{C}_{s,\,p}$, is

$$\mathbf{C}_{s,P} = \begin{bmatrix} \mathbf{C}_{i1} & 0 & \dots & 0 \\ 0 & \mathbf{C}_{i2} & \dots & 0 \\ \dots & \dots & \dots & \dots \\ 0 & 0 & \dots & \mathbf{C}_{iP} \end{bmatrix} \tag{6.76}$$

That is, it is a diagonal matrix of submatrices, where each submatrix is the noise correlation matrix of one source.

An implicit assumption of Section 6.2.3 is that the modulated noise sources in the circuit are themselves uncorrelated; that is, all correlations are between sidebands, not between the various sources. This assumption simplifies the analysis, and there is justification for it. In linear noise analysis, we encounter correlated sources. Those sources, however, invariably represent short-circuit noise currents (or open-circuit noise voltages) at ports or nodes, which result from the circuit's internal noise sources. Because the currents at a pair of ports or nodes arise, to some degree, from the same sources, they are correlated. The internal sources, however, which generate those port currents, are rarely correlated. In nonlinear noise analysis, we deal directly with the internal noise sources, and, therefore, only rarely encounter correlations between them.

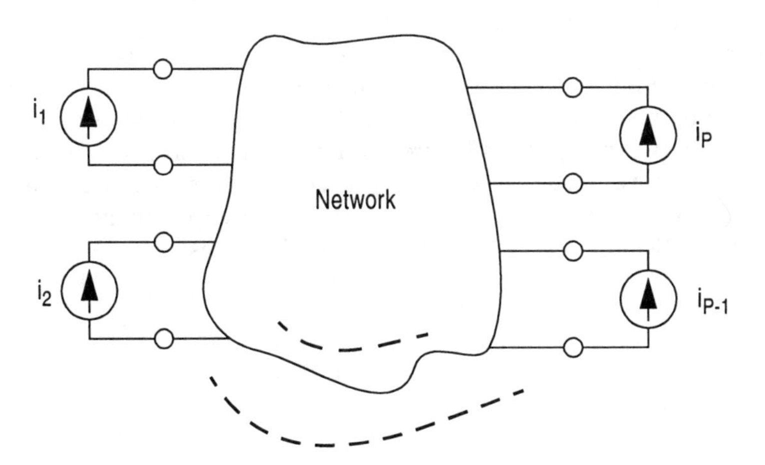

Figure 6.4 Illustration of the noise-analysis problem. The network is a linearized, time-varying nonlinear circuit, excited by the circuit elements' noise sources, a collection of both modulated and unmodulated sources.

We do not extend this assumption to linear sources. In many linear sources, such as lossy transmission lines, the noise-generating elements are distributed, so we cannot treat them as distinct sources. The only valid characterization is to calculate a noise-correlation matrix in some other way. For a passive, lossy element, that calculation is simple; the current noise-correlation matrix is calculated directly from its admittance matrix (Section 5.3.2.2). To perform a noise analysis of a nonlinear microwave circuit, we must include such sources. Doing so is straightforward.

The goal of the nonlinear noise analysis is to obtain the mean-square noise voltage at the component's output port. Then the noise temperature can be determined by methods given in Chapters 3 and 5. To do this, we need to obtain a reduced voltage noise-correlation matrix. The correlation matrix also could be used to determine a set of noise parameters, in the same manner as with a linear circuit. We could use such parameters, in principle, to calculate the noise parameters of a nonlinear two-port, such as a mixer, in precisely the same manner as was used for linear two-ports. This formulation is less useful for nonlinear circuits, however, because any attempt to noise-match the nonlinear circuit would affect its embedding impedances at mixing frequencies other than the input, and would therefore change the noise parameters! In fact, little would be gained by such efforts. In linear circuits, the noise parameters of a transistor allow us to determine the device's noise under arbitrary matching conditions, and we can adjust the matching without affecting the device's noise parameters. This is not the case in pumped nonlinear circuits. Therefore, we usually must treat the matching circuits in nonlinear components as an integral part of the component and calculate the noise temperature with those circuits in place.

Because our formulation is fundamentally nodal, we must convert the correlation matrix to a nodal form. To do this, we modify the circuit to create the equivalent shown in Figure 6.5. Each source has been converted to a pair of sources, in such a way that it is equivalent to the circuit in Figure 6.4. This operation is an extension of the one used for unmodulated noise sources in Section 5.3.2.3. The noise vector becomes

$$\mathbf{I}_n = \begin{bmatrix} \mathbf{i}_{n1} \\ \mathbf{i}_{n2} \\ \mathbf{i}_{n3} \\ \mathbf{i}_{n4} \\ \cdots \\ \mathbf{i}_{n2P-1} \\ \mathbf{i}_{n2P} \end{bmatrix} = \begin{bmatrix} \mathbf{i}_1 \\ -\mathbf{i}_1 \\ \mathbf{i}_2 \\ -\mathbf{i}_2 \\ \cdots \\ \mathbf{i}_P \\ -\mathbf{i}_P \end{bmatrix} \tag{6.77}$$

where $\mathbf{i}_{np}$ are vectors of the nodal currents at the ports p in the form of (6.71) and $\mathbf{i}_p$ are the vectors of noise-source currents in the same form. The elements of $\mathbf{I}_n$ are a combination, in general, of modulated and unmodulated noise sources. Some nodes, of course, may not have noise sources connected to them, so the corresponding current vectors in (6.77) are zero.

The resulting correlation matrix for the nodal currents, $C_{s,\,N}$, is

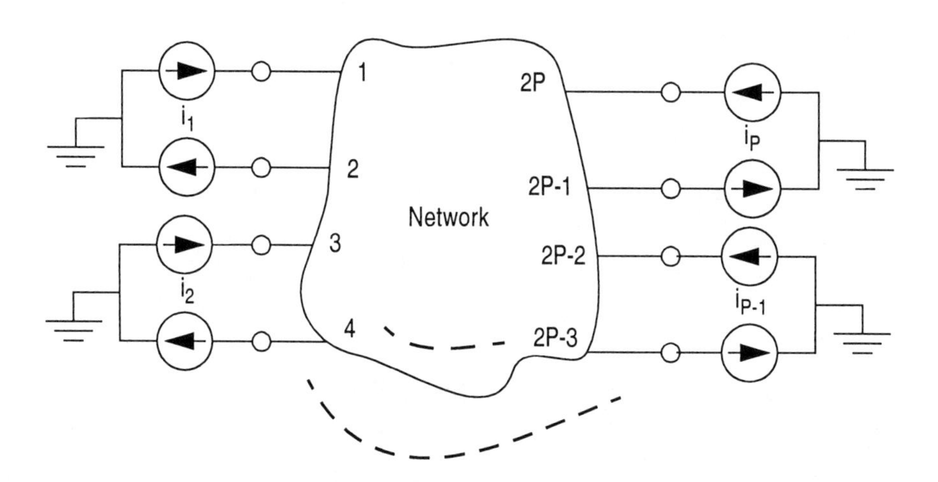

Figure 6.5 The circuit of Figure 6.4 can be converted into the form shown here, allowing for nodal analysis. The P ports, to which noise sources are connected, are converted to 2P nodes.

$$\mathbf{C}_{s,N} = \mathbf{I}_n\mathbf{I}_n^{*T} = \begin{bmatrix} \mathbf{C}_{i1} & -\mathbf{C}_{i1} & 0 & 0 & \dots & 0 & 0 \\ -\mathbf{C}_{i1} & \mathbf{C}_{i1} & 0 & 0 & \dots & 0 & 0 \\ 0 & 0 & \mathbf{C}_{i2} & -\mathbf{C}_{i2} & \dots & 0 & 0 \\ 0 & 0 & -\mathbf{C}_{i2} & \mathbf{C}_{i2} & \dots & 0 & 0 \\ \dots & \dots & \dots & \dots & \dots & \dots & \dots \\ 0 & 0 & 0 & 0 & \dots & \mathbf{C}_{iP} & -\mathbf{C}_{iP} \\ 0 & 0 & 0 & 0 & \dots & -\mathbf{C}_{iP} & \mathbf{C}_{iP} \end{bmatrix} \tag{6.78}$$

where the $\mathbf{C}_{ip}$ are the elements of (6.76).

Finally, we need to add the noise of the linear parts of the circuit to the correlation matrix. If the linear element is a simple two-terminal passive admittance, its noise can be included in the same manner as the modulated sources. The correlation matrix for such admittances is simply a diagonal. From (6.74) and (6.75),

$$\mathbf{C}_{ip} = 4KT_0\Delta f \operatorname{diag}(\operatorname{Re}\{Y(\omega_k)\}) \tag{6.79}$$

This is, of course, the port formulation, and must be copied into four places in the $\mathbf{C}_{s,N}$ matrix as shown in (6.78). The inclusion of larger sets of sources creates off-diagonal terms in the submatrices of $\mathbf{C}_{s,N}$. Suppose, for example, we have correlated linear sources at nodes p and q. The correlation matrix, at frequency ω_k, is

$$\mathbf{C}_{ip,q;k} = \left.\begin{bmatrix} \overline{i_p i_p^*} & \overline{i_p i_q^*} \\ \overline{i_q i_p^*} & \overline{i_q i_q^*} \end{bmatrix}\right|_{\omega_k} \tag{6.80}$$

The inclusion of these terms is most simply shown by beginning with the port-form noise correlation matrix, (6.76), and converting it to a nodal matrix. The terms are added to the (p, p), (p, q), (q, p), and (q, q) positions in the locations of the submatrices corresponding to ω_k:

$$
\mathbf{C}_{s,P} = \begin{array}{cc}
 & \begin{array}{ccc} \quad p & \quad \cdots \quad & q \quad \end{array} \\
\begin{array}{c} p \\ \vdots \\ \\ \vdots \\ q \end{array} &
\begin{bmatrix}
\ddots & & & \\
 & \begin{bmatrix} \ddots & & \\ & \overline{i_p i_p^*}\big|_{\omega_k} & \\ & & \ddots \end{bmatrix} & & \begin{bmatrix} \ddots & & \\ & \overline{i_p i_q^*}\big|_{\omega_k} & \\ & & \ddots \end{bmatrix} \\
 & & \ddots & \\
 & \begin{bmatrix} \ddots & & \\ & \overline{i_q i_p^*}\big|_{\omega_k} & \\ & & \ddots \end{bmatrix} & & \begin{bmatrix} \ddots & & \\ & \overline{i_q i_q^*}\big|_{\omega_k} & \\ & & \ddots \end{bmatrix}
\end{bmatrix}
\end{array}
\tag{6.81}
$$

The $\overline{i_p i_q^*}$ and $\overline{i_q i_p^*}$ terms form off-diagonal submatrices in (6.76), and all four submatrices are themselves diagonals. In general, there is a matrix of the form (6.80) for each ω_k. Of course, terms in the $-k$ positions represent negative frequencies, and thus must be conjugate. To create the nodal-form matrix, the submatrices of (6.81) are simply copied into four places as was done with the $\mathbf{C}_{ip}$ matrices in (6.78).

Similarly, we saw in Section 5.3.2.2 that the current noise correlation matrix of a passive circuit, at some frequency ω, is

$$\mathbf{C}_i = 4KT\Delta f \,\mathrm{Re}\{\mathbf{Y}(\omega)\} \tag{6.82}$$

so we add $4KT\Delta f \,\mathrm{Re}\{Y_{p,q}(\omega_k)\}$ to all (p, q) positions of corresponding to ω_k. This effectively adds the diagonal matrices

$$\mathbf{C}_{ip,q} = 4KT_0\Delta f \mathrm{diag}(\mathrm{Re}\{Y_{p,q}(\omega_k)\}) \tag{6.83}$$

to all (p, q) positions, as shown in (6.81). Again, we convert the port correlation matrix to a nodal correlation matrix as we did for the case of the pair of correlated sources.

6.3.2 Noise Analysis

Our goal is to determine the mean-square noise voltage at the output port, which is an element of the voltage noise correlation matrix. Knowing this voltage, we can easily determine the noise output power, and thus the equivalent input noise temperature, however one wishes to define it (Section 3.2.1).

Since our noise excitations are currents, we must obtain the voltage correlation matrix. As before,

$$\mathbf{C}_i = \mathbf{Y}_c\mathbf{C}_v\mathbf{Y}_c^{*\mathrm{T}} \tag{6.84}$$

is solved to obtain $\mathbf{C}_v$, the voltage noise correlation matrix. $\mathbf{Y}_c$ is the conversion matrix of the network. The simplest solution, in concept, is to invert $\mathbf{Y}_c$ and form

$$\mathbf{C}_v = \mathbf{Z}_c\mathbf{C}_i\mathbf{Z}_c^{*\mathrm{T}} \tag{6.85}$$

where

$$\mathbf{Z}_c = \mathbf{Y}_c^{-1} \tag{6.86}$$

We have noted earlier that this process, while intuitively satisfying, is numerically inefficient. The method described in Section 5.3.2.4 for solving equations of the form (6.85) should be applied here as well.

Once we have $\mathbf{C}_v$, reducing the matrix to a two-port is straightforward. There is little difference, mathematically, between the reduction for the nonlinear case and the linear case, described in Section 5.3.2.5. The only difference is that we must select the port voltage components at the correct frequencies. Following the same procedure as in Section 5.3.2.5, we have

$$\begin{aligned} v_1 &= v_m - v_n\big|_{\omega_j} \\ v_2 &= v_p - v_q\big|_{\omega_k} \end{aligned} \tag{6.87}$$

where the nodes (m, n) are the input port, port 1, and (p, q) are the output, port 2. ω_j is the input frequency and ω_k is the output. As before,

$$\begin{aligned}
\overline{v_1 v_1^*} &= \overline{|v_1|^2} = \overline{|v_m|^2} + \overline{|v_n|^2} - 2Re\{\overline{v_m v_n^*}\} \\
\overline{v_2 v_2^*} &= \overline{|v_2|^2} = \overline{|v_p|^2} + \overline{|v_q|^2} - 2Re\{\overline{v_p v_q^*}\} \\
\overline{v_1 v_2^*} &= \overline{v_m v_p^*} + \overline{v_n v_q^*} - \overline{v_n v_p^*} - \overline{v_m v_q^*}
\end{aligned} \tag{6.88}$$

where the voltage components v_m, v_n, v_p, and v_q are taken at their appropriate frequencies. We now have a 2×2 voltage noise correlation conversion matrix. The corresponding current noise correlation matrix is

$$\mathbf{C}_i = \mathbf{Y}\mathbf{C}_v\mathbf{Y}^{*\mathrm{T}} \tag{6.89}$$

where $\mathbf{C}_v$, now a 2×2 matrix, has the components given in (6.88). **Y** is now the 2×2 reduced admittance matrix representing the two-port, with the input at ω_j and output at ω_k, which is generated in the manner shown in Section 6.1.4. The most important term in this matrix is the mean-square output voltage, $\overline{v_2 v_2^*} = \overline{|v_2|^2}$, which gives us the output noise temperature. The equivalent input noise temperature is the output noise temperature divided by the transducer gain. Having $\mathbf{C}_i$, we could also find $F_{\min}$, R_n, and $Y_{s,\,\mathrm{opt}}$, if desired, from the expressions in Section 5.2.4.

It is worthwhile to note that, once we account for modulated sources, there is little fundamental difference between the mathematical treatments of the linear and nonlinear cases. The situation is analogous to the treatment of admittance matrices of linear circuits and conversion matrix analysis of linearized time-varying circuits.

6.4 NOISE IN AUTONOMOUS CIRCUITS

6.4.1 Autonomous Circuits

We now consider *autonomous* circuits, ones that have no external RF excitation. (They do have dc bias, however.) In practice, sinusoidal oscillators are the only autonomous circuit we regularly deal with. Although much of the theory in Section 6.2 is useful in the analysis of oscillators, such circuits have a special set of problems that we must address.

As with nonautonomous circuits, our approach is to perform a nonlinear analysis, linearize it, and perform the noise analysis of the linearized, time-varying circuit. The first problem involves the nonlinear analysis, where we face three fundamental difficulties. First, all oscillatory circuits have a *zero solution*; that is, a nonoscillating state satisfies the circuit equations. Unless we do something special to avoid it, a harmonic-balance analysis of an oscillator (as well as a time-domain analysis) invariably finds this zero solution and blithely terminates the solution process. In effect, we are dealing with a circuit that, by its very nature, has multiple solutions. Circuit simulators do not deal with such circuits very well, partly because of their inability to read the user's mind and determine which solution he is interested in.[3] Second, the initial phase of the solution is indeterminate. In nonautonomous circuits, the phase is defined by the excitation, but in oscillators, any value of phase is acceptable; there is no inherent zero value of time. This lack of a time reference and the resulting indeterminacy of the solution implies that the circuit equations are ill conditioned. Finally, harmonic-balance analysis requires that the frequencies in the circuit be specified. In nonautonomous circuits, the mixing frequencies are defined by the excitations, but in an oscillator, the frequency of oscillation is unknown a priori. Determining the frequency must be part of the analysis.

Various methods for performing a harmonic-balance analysis of an oscillator have been developed [6.3–6.7]. Oscillators can be analyzed by time-domain techniques as well as harmonic balance. In time-domain analysis, a transient of some type, such as an impulse or dc turn-on, is used to start the oscillation and avoid the zero solution. As the oscillation builds, it creates its own time reference, ultimately introduced by the turn-on transient, and the oscillation relaxes to its steady-state conditions. Thus, the latter two problems—time reference and frequency—are not significant problems in time-domain analysis. The transient source can also be helpful in avoiding the zero solution. An analogous treatment, using a so-called *auxiliary generator*, can be used in harmonic-balance analysis.

The second problem is in performing the noise analysis itself. The conversion matrix of any nonlinear circuit is simply a Jacobian matrix that has been offset in frequency from the excitation or, in the case of oscillators, the frequency of oscillation. The Jacobian of an autonomous circuit, unfortunately, is singular, so the conversion matrix is ill conditioned at small offsets. Unfortunately, the small offsets are frequently of greatest interest. Because of the ill conditioning, conversion matrix noise analysis, as described in Section 6.3.2, has met with, at best, mixed success when applied to oscillators. As a result, oscillator noise has been formulated in other

---

3. In fact, one of the fundamental characteristics of nonlinear functions is the presence of multiple solutions. Thus, many types of both autonomous and nonautonomous nonlinear circuits have multiple modes of operation. Usually those "spurious" solutions are far removed from the expected one, so, in practice, the desired solution usually is found.

ways, which have generally been more successful than the conversion matrix formulation.

6.4.2 Modulation and Conversion Noise

In [6.8, 6.9] Rizzoli et al. introduced the concepts of *modulation noise* and *conversion noise*. Conversion noise analysis is the direct application of the methods of Section 6.2 to oscillator noise. The method works acceptably for the analysis of noise at large offsets from the carrier, but not well for *close-in noise*, that is, noise close to the carrier. To circumvent this difficulty, they proposed an alternate approach, which they called *modulation noise*. In this case, the noise was treated as a phase perturbation of the oscillatory steady state. This approach worked much better for close-in noise.

It has always been a little disturbing that these equivalent approaches to noise analysis do not give the same result. The problem is not a lack of equivalence between the conversion and modulation approaches, but in the numerical characteristics of the resulting equations. We shall see examples in Chapter 8 showing that, for small frequency offsets, the conversion formulation is severely ill conditioned. It is possible to show, in special cases that can be controlled adequately, correct results for near-carrier noise, obtained with the conversion formulation [6.4, 6.10, 6.11]. Even in such cases, however, small changes in parameters cause large changes in the noise spectrum. (See Section 8.3.4 for an example.) In more general, practical problems, the conversion approach is far too badly conditioned to provide acceptable near-carrier results.

6.4.3 Conversion Noise

6.4.3.1 Noise Analysis

The simplest approach to noise analysis of autonomous circuits is to use the noise theory developed for nonautonomous circuits [6.12]. In this process, the conversion matrix of the circuit at the offset frequency of interest, $\mathbf{Y}_c$, is formulated in the manner described in Section 6.1.4. We shall assume that this matrix is in nodal form. The noise voltage at the circuit's nodes is found by solving

$$\mathbf{I}_n = \mathbf{Y}_c \mathbf{V}_n \tag{6.90}$$

where $\mathbf{I}_n$ and $\mathbf{V}_n$ are, respectively, the vectors of noise excitation currents and voltages at all sidebands and have the form of (6.35). In this section, as previously, we assume that all noise sources are current sources; the extension to a more general case, if warranted, is straightforward. The voltage and current noise correlation matrices are related, as before, by

$$\mathbf{C}_i = \mathbf{I}_n\mathbf{I}_n^{*T} = \mathbf{Y}_c\mathbf{V}_n\mathbf{V}_n^{*T}\mathbf{Y}_c^{*T} = \mathbf{Y}_c\mathbf{C}_v\mathbf{Y}_c^{*T} \tag{6.91}$$

and (6.91) can be solved to obtain $\mathbf{C}_v$. The mean-square noise voltages in $\mathbf{C}_v$ represent a combination of amplitude and phase noise; in an oscillator, it is a fair (and common) assumption that phase noise dominates.

The conversion matrix, $\mathbf{Y}_C$, is a Jacobian matrix evaluated at the offset frequency of the noise. That offset frequency can be as small as a few hertz, making the conversion matrix insignificantly different from the zero-offset Jacobian. The latter is theoretically singular, so ill conditioning of the conversion matrix should be expected. Bolcato et al. [6.11] make the point that this problem can be controlled if the circuit equations for the oscillator, especially the determination of the frequency of oscillation, are solved to a high degree of precision. Most recent work on oscillator phase noise, however, has centered on a modulation noise approach, which is, at least, considerably easier to make successful.

6.4.3.2 Phase and Amplitude Components

The noise analysis, however performed, results in a noise spectral density. That noise consists of a combination of phase and amplitude components, only one of which may affect a particular type of communication system. Thus, it is important to be able to separate the two.

$\mathbf{C}_v$, as before, is a matrix of submatrices. That is, it describes pairs of noise sidebands and their correlations, centered on each harmonic of the oscillator frequency, as shown in Figure 6.2. Both sidebands contribute to the phase and amplitude noise at any harmonic. In general, the sidebands are correlated, and we must include the effects of their correlation.

The components on the main diagonal of the submatrix corresponding to the port of interest are the mean-square noise-voltage sidebands at each of the harmonics. The off-diagonal elements give the correlations between different frequency components. Although the fundamental frequency is usually the only one of interest, in some types of oscillators, such as a frequency-doubling oscillator, the phase noise at a harmonic may be of primary interest instead. We can view any harmonic and its noise sidebands as sinusoids that vary slowly relative to each other. The resultant is a sine wave, the oscillator output, that varies slowly in phase and amplitude. In this sense, the situation is identical to a carrier and modulation sidebands.

As with a modulated carrier, the components of the modulation/noise in phase with the carrier contribute to amplitude variations, while the components perpendicular to it contribute to phase variations.

We first examine the phase variations. The carrier phasor, V_c, is

$$V_c = V_k \exp(j(k\omega_p + \phi_k)) \tag{6.92}$$

where V_k is the magnitude of the carrier component at the kth harmonic. The upper and lower modulation sideband components perpendicular to it, v_u and v_l, are

$$\begin{aligned} v_u &= v_{ku} \exp\left(j\left(\phi_k + \frac{\pi}{2}\right)\right) \\ v_l &= v_{kl} \exp\left(-j\left(\phi_k + \frac{\pi}{2}\right)\right) \end{aligned} \tag{6.93}$$

where v_{ku} and v_{kl} are phasors at the kth harmonic. The phase of the lower sideband is negative because it is counter-rotating, relative to the upper. The phase deviation at any instant is

$$\Delta\phi = \operatorname{atan}\left(\frac{|v_u + v_l|}{|V_c|}\right) \approx \frac{|v_u + v_l|}{|V_c|} \tag{6.94}$$

since $|v_u + v_l| \ll |V_c|$. The mean-square phase deviation is, then,

$$\overline{|\Delta\phi|^2} = \frac{\overline{|v_u + v_l|^2}}{|V_c|^2} \tag{6.95}$$

and substituting (6.92) and (6.93) into (6.95) gives

$$\overline{|\Delta\phi|^2} = \frac{\overline{|v_{ku}|^2} + \overline{|v_{kl}|^2} - 2\mathrm{Re}\{\overline{v_{ku} v_{kl}^*} \exp(2j\phi_k)\}}{V_k^2} \tag{6.96}$$

which agrees, in the limit, with simple analyses based on zero correlation between the sidebands [6.13].

The in-phase sideband components are

$$\begin{aligned} v_u &= v_{ku}\exp(j\phi_k) \\ v_l &= v_{kl}\exp(-j\phi_k) \end{aligned} \tag{6.97}$$

and the amplitude noise is

$$\overline{|v_n|^2} = \overline{|v_u + v_l|^2} \tag{6.98}$$

Substituting (6.97) into (6.98) gives

$$\overline{|v_n|^2} = \overline{|v_{ku}|^2} + \overline{|v_{kl}|^2} + 2\mathrm{Re}\{\overline{v_{ku}v_{kl}^*}\exp(2j\phi_k)\} \tag{6.99}$$

If $\overline{|v_n|^2}$ is a mean-square output noise voltage, it can be converted into a noise temperature, if desired.

6.4.4 Modulation Noise

We have made the point that ill conditioning of the Jacobian matrix limits the numerical performance of conventional nonlinear noise analysis—conversion noise analysis—in autonomous circuits. This ill conditioning results, in part, from the indeterminacy of the solution. Removing this indeterminacy, by introducing an absolute time reference (say, by constraining a frequency component's phase to zero), does not entirely solve the problem; the Jacobian, in an autonomous circuit, is still ill conditioned. One solution, called *modulation-noise analysis*, is to include frequency or phase fluctuations explicitly. The result is an approximation that works well at small frequency offsets. The conversion noise approach is valid at large offsets, so a combination of the two methods should be successful for the general case.

The method is easily illustrated. For simplicity, we consider the same case as in conversion noise, where the noise is generated entirely by current noise sources. The extension to the general case is straightforward.

In the conversion-noise approach, we employ a conversion matrix, which is nothing more than a Jacobian matrix evaluated at the offset frequency of the noise. The noise currents and voltages, at all the sidebands, are related as

$$\mathbf{I}_n = \mathbf{Y}_c \mathbf{V}_n \tag{6.100}$$

In this view, the sideband noise is converted to other sidebands through a frequency conversion process, not unlike what occurs in a mixer. In the modulation-noise view, we treat the noise excitations as a perturbation, at the oscillation frequency, of the large-signal waveforms.

This process depends upon the use of a frequency-component replacement, which will be described further in Section 8.2.2.2. Briefly, in the harmonic-balance analysis of the oscillator, we substitute the oscillation frequency for the imaginary part of the fundamental frequency. Eliminating the imaginary part provides a phase reference, and replacing it with the oscillator frequency allows the frequency to be determined naturally as part of the ordinary harmonic-balance iterations. The Jacobian is formulated accordingly; we have

$$\begin{bmatrix} \delta I_0 \\ \delta I_{1r} \\ \delta I_{1i} \\ \delta I_{2r} \\ \delta I_{2i} \\ \cdots \end{bmatrix} = \begin{bmatrix} \frac{\partial I_0}{\partial V_0} & \frac{\partial I_0}{\partial V_{1r}} & \frac{\partial I_0}{\partial \omega} & \frac{\partial I_0}{\partial V_{2r}} & \cdots & \cdots \\ \frac{\partial I_{1r}}{\partial V_0} & \frac{\partial I_{1r}}{\partial V_{1r}} & \frac{\partial I_{1r}}{\partial \omega} & \frac{\partial I_{1r}}{\partial V_{2r}} & \cdots & \cdots \\ \frac{\partial I_{1i}}{\partial V_0} & \frac{\partial I_{1i}}{\partial V_{1r}} & \frac{\partial I_{1i}}{\partial \omega} & \cdots & \cdots & \cdots \\ \cdots & \cdots & \cdots & \cdots & \cdots & \cdots \\ \cdots & \cdots & \cdots & \cdots & \cdots & \cdots \\ \cdots & \cdots & \cdots & \cdots & \cdots & \cdots \end{bmatrix} \begin{bmatrix} \delta V_0 \\ \delta V_{1r} \\ \delta \omega \\ \delta V_{2r} \\ \delta V_{2i} \\ \cdots \end{bmatrix} \tag{6.101}$$

where δI_{kr} and δV_{kr} are perturbations of the real part of the current and voltage, respectively, at the *k*th harmonic, and δI_{ki} and δV_{ki} are the imaginary parts. To keep the illustration clear, (6.101) is formulated for only a single port; it is a simple matter to extend it to the general harmonic-balance case. The Jacobian in (6.101) clearly contains information about the variation of frequency caused by each noise component. The voltage vector, including the frequency perturbation, $\delta\omega$, is

$$\delta \mathbf{V} = \mathbf{J}^{-1} \delta \mathbf{I} \tag{6.102}$$

where $\mathbf{J}$ is the Jacobian matrix of (6.101) and $\delta\mathbf{V}$ and $\delta\mathbf{I}$ are the vectors. Then, $\delta\omega$ is given by

$$\delta\omega = \mathbf{J}_r^i \delta\mathbf{I} \tag{6.103}$$

where $\mathbf{J}_r^i$ is the appropriate row of the inverse Jacobian.

Equations (6.101) to (6.103) are valid only for small sinusoidal perturbations. The noise, however, consists of modulation components above and below each large-signal harmonic. To make the relation valid for noise, we must replace the sinusoids with noise quantities, consistent with modulation laws. The noise vector becomes

$$\mathbf{I}_n = \begin{bmatrix} i_{n,0} \\ i_{n,1r} + i_{n,-1r} \\ -i_{n,1i} + i_{n,-1i} \\ \cdots \\ \cdots \end{bmatrix} \tag{6.104}$$

where the subscripts r and i indicate the real and imaginary components, respectively. We have assumed that the Jacobian formulation allows the real and imaginary parts of the current to be separated as shown. This is a common practice in computer-aided circuit analysis.

The mean-square frequency deviation is

$$\overline{|\delta\omega|^2} = \overline{\delta\omega\delta\omega^*} = \mathbf{J}_r\mathbf{I}_n\mathbf{I}_n^{*\mathrm{T}}\mathbf{J}_r^{*\mathrm{T}} = \mathbf{J}_r\mathbf{C}_i\mathbf{J}_r^{*\mathrm{T}} \tag{6.105}$$

and, from the theory of ordinary frequency and phase modulation, the mean-square phase deviation is simply

$$\overline{|\delta\phi|^2} = \frac{k^2}{\omega^2}\overline{|\delta\omega|^2} \tag{6.106}$$

In virtually all oscillators, the phase noise is dominated by $1/f$ noise in the solid-state device. In such cases, the phase noise spectrum varies 30 dB per decade of offset frequency. The failure of conversion-noise analysis is evident as a phase-noise spectrum that does not vary as expected at small offsets; typically, the slope is flatter than expected. Modulation noise increases as expected at small offsets, but steadfastly retains its 30 dB per octave slope well beyond the point where, in real oscillators, the noise spectrum flattens. Clearly, at some point the two curves must cross over, and the treatment of noise in this region is somewhat problematical. Technically, the two noise processes should be combined, with due regard to their correlation properties. In practice, however, one simply switches between formulations at some appropriate offset frequency. To do this, of course, it is necessary to analyze the oscillator noise according to both the conversion and modulation theories in any single analysis.

6.4.5 Envelope Noise Analysis

Ngoya et al. [6.4] proposed an approach to noise analysis that avoids both the approximations of modulation noise that lead to errors at large offsets and the ill conditioning of conversion noise. It is based on *envelope analysis*, sometimes called *envelope transient analysis*.

6.4.5.1 Envelope Analysis

Envelope analysis is a method that appears to have been invented several times [6.14–6.16]. Related to harmonic-balance analysis, it is useful for analyses involving modulated signals or, in this case, noise. It involves replacing the multitone harmonic-balance analysis with sequential single-tone harmonic-balance analyses. The time intervals are based on sampling at the envelope frequency, not at the carrier frequency.

The claim has been made that envelope analysis is fundamentally more efficient than multitone harmonic-balance analysis. This claim is true, but the advantage is not as great as one might expect. It is possible to select time intervals for multitone harmonic balance based on a two-dimensional sampling, which is equivalent, in some sense, to the envelope-analysis case. The real advantage, in terms of computational speed, comes from the way the computational effort scales with matrix size. The time required to factor the Jacobian matrix increases more than linearly with its size, so it is better to factor a smaller matrix more often than a large matrix less often. Envelope analysis favors the sequential factoring of a smaller matrix. For a more concrete view, suppose that a single-tone analysis required a Jacobian of dimension K, while requiring N samples over the envelope period. An equivalent multitone analysis would require a Jacobian of dimension KN, which requires more computational effort than solving a K-dimension matrix N times. However, modern matrix methods can approach linear scaling, and envelope analysis carries with it other forms of computational overhead that are missing from multitone harmonic balance. These significantly reduce the speed advantage of envelope analysis.

A more important characteristic of envelope analysis is its ability to handle nonperiodic or stochastic signals. This allows components to be analyzed with real signals instead of periodic approximations. Then, the method can directly calculate quantities, such as bit-error rates, which cannot be determined easily from sinusoidal analyses.

In envelope analysis, all voltages, currents, charges, and similar quantities are treated as narrowband spectra in the vicinity of each large-signal harmonic. In effect, each harmonic has a modulation associated with it. All quantities can be expressed as

$$x(t) = \sum_{k=-K}^{K} X_k(t) \exp(jk\omega_p t) \tag{6.107}$$

where $X_k(t)$ is a modulating function in the vicinity of the frequency $m\omega_p$. For a periodic $X_k(t)$,

$$X_k(t) = \sum_{n=-N}^{N} X_n \exp(jn\omega_0 t) \tag{6.108}$$

and ω_0 is the fundamental modulation frequency. We assume that $X_k(t)$ is narrowband; $\omega_0 \ll \omega_p$. This guarantees that $X_k(t)$ varies slowly relative to the harmonic, so the harmonic balance simulation can be carried out at a sequence of time points based on the envelope period (or inverse bandwidth, in the case of nonperiodic signals). At each time interval, $X_k(t)$ is sampled and a harmonic-balance analysis performed.

There are a couple of complications in this process. First, frequently we must differentiate a charge waveform to obtain current. The charge waveform, $Q(t)$, is

$$Q(t) = \sum_{k=-K}^{K} Q_k(t) \exp(jk\omega_p t) \tag{6.109}$$

The current, $I_q(t)$, is

$$
\begin{aligned}
I_q(t) &= \frac{d}{dt}Q(t) \\
&= \sum_{k=-K}^{K} \left(jk\,\omega_p Q_k(t) + \frac{d}{dt}Q_k(t) \right) \exp(jk\omega_p t)
\end{aligned}
\quad (6.110)
$$

which includes a term $dQ_k(t)/dt$, which does not appear in conventional harmonic balance analysis.

Second, we must account for the variation of the linear subcircuit's admittance over the modulation frequency at each harmonic. A simple method is to linearize the admittance function near the harmonic frequency:

$$
Y(k\omega_p + \delta\omega) = Y(k\omega_p) + \left.\frac{dY(\omega)}{d\omega}\right|_{\omega = k\omega_p} \delta\omega \quad (6.111)
$$

where Y represents any admittance term in (6.4).

6.4.5.2 Noise Analysis

We now show how the envelope concepts in Section 6.4.5.1 can be applied to noise analysis. For simplicity, we assume that we have a single nonlinear element. The extension to multiple nonlinear elements is straightforward; it involves adding voltage and current vectors, for each port, as shown in (6.1) and (6.2).

Expressed in the time domain, the circuit equation is

$$
i_L(t) + \frac{d}{dt}q[v(t)] + i_{NL}(t) + i_s(t) = 0 \quad (6.112)
$$

where i_L is the current in the linear subcircuit, i_{NL} is the current in the nonlinear resistive elements, q is the nonlinear charge function, and i_s is the excitation. We assume that all quantities have the form,

$$
x(t) = \sum_{k=-K}^{K} X_k(t) \exp\left(jk\omega_p t + jk \int_{-\infty}^{t} \delta\omega(\tau)\,d\tau \right) \quad (6.113)
$$

where $x(t)$ can represent voltage, current, or charge. In effect, we have added a phase deviation to the modulating waveform, $X_k(t)$. It is axiomatic that all voltages and currents, and functions of those quantities, must have the same frequency, ω_p, and frequency deviation, $\delta\omega$.

We wish to form an expression in the frequency domain, in terms of the envelope terms $X_k(t)$. We then perform a perturbation analysis, not unlike the conversion-matrix analysis, but use it to determine $\delta\omega$ as well as the small-signal voltages and currents. The inclusion of the frequency term prevents the ill conditioning that we would encounter in classical conversion-matrix analysis.

We substitute voltage, in the form of (6.113), into (6.112). The only complication is the charge term. We have

$$q(t) = \sum_{k=-K}^{K} Q_k(t) \exp\left(jk\omega_p t + jk\int_{-\infty}^{t} \delta\omega(\tau)d\tau \right) \tag{6.114}$$

so

$$\frac{d}{dt}q(t) = \sum_{k=-K}^{K} \left(Q_k(t)\,(jk\omega_p + jk\delta\omega(t)) + \frac{d}{dt}Q_k(t) \right) \cdot \exp\left(jk\omega_p t + jk\int_{-\infty}^{t} \delta\omega(\tau)d\tau \right) \tag{6.115}$$

After we drop the clumsy exponential term in (6.113) to (6.115), (6.112) becomes

$$\mathbf{YV}(t) + \frac{d}{dt}\mathbf{Q}(t) + j\Omega\mathbf{Q}(t) + \mathbf{I}_{NL}(\mathbf{V}) + \mathbf{I}_s = 0 \tag{6.116}$$

where

$$\mathbf{V} = \begin{bmatrix} V_{-M}(t) \\ V_{-M+1}(t) \\ \dots \\ V_0(t) \\ V_1(t) \\ \dots \\ V_M(t) \end{bmatrix} \tag{6.117}$$

and

$$\begin{aligned} \Omega &= \mathrm{diag}(k\omega_p + k\delta\omega) \qquad k = -K\ldots K \\ &= \Omega_p + \delta\Omega \end{aligned} \tag{6.118}$$

Note that ω_p is a known quantity, the result of the oscillator's harmonic-balance analysis, but $\delta\omega$ is a variable, which we must determine. This variable increases the number of variables to one more than the number of equations; to resolve the matter, we note that the phase of one term must be set to zero, to create a phase reference. The logical choice is to set the imaginary part of the fundamental-frequency voltage component to zero:

$$V_1(t) + V_{-1}(t) = 0 \tag{6.119}$$

This additional constraint equalizes the number of variables and equations.

We now perturb the voltage and extract the perturbations of the dependent quantities. Then,

$$\mathbf{V} = \mathbf{V}_0 + \delta\mathbf{V} \tag{6.120}$$

where $\delta\mathbf{V}$ is the perturbation, which, in general, is at an offset frequency, and $\mathbf{V}_0$ is the unperturbed, steady-state response of the oscillator. The charge is

$$\mathbf{Q}(\mathbf{V}_0 + \delta\mathbf{V}) = \mathbf{Q}(\mathbf{V}_0) + \frac{d\mathbf{Q}}{d\mathbf{V}}\delta\mathbf{V} \tag{6.121}$$

$\mathbf{Q}(\mathbf{V}_0)$ is the steady-state charge and $d\mathbf{Q}/d\mathbf{V}$ is a conversion matrix. A similar expression applies to the nonlinear-current term.

We now substitute (6.120) and (6.121) into (6.116) and subtract the steady state to obtain the equation for the perturbation. We obtain

$$\mathbf{Y}\delta\mathbf{V} + \frac{d}{dt}\left(\frac{d\mathbf{Q}}{d\mathbf{V}}\delta\mathbf{V}\right) + j\Omega_p\frac{d\mathbf{Q}}{d\mathbf{V}}\delta\mathbf{V} + j\delta\Omega\mathbf{Q} + \frac{d\mathbf{I}_{NL}}{d\mathbf{V}}\delta\mathbf{V} + \delta\mathbf{I}_s = 0 \qquad (6.122)$$

Recall that

$$\delta\Omega = \delta\omega\,\mathrm{diag}(k) \qquad k = -K\ldots K \qquad (6.123)$$

so

$$j\delta\Omega\mathbf{Q} = j\mathbf{D}_M\mathbf{Q}\delta\omega \qquad (6.124)$$

where $\mathbf{D}_M$ is the diagonal matrix in (6.123).

$\delta\mathbf{I}_s$ is, of course, the noise excitation. In general, noise excitations may be significant only near a single harmonic (such as $1/f$ noise, which exists only at baseband) or at all harmonics (device high-frequency noise and thermal noise). Through the use of modulation laws [6.8], we have

$$\delta I_{s,k} = \frac{1}{2}[N_k^+\exp(j(k\omega_p + \omega_0)) + N_k^-\exp(j(k\omega_p - \omega_0))] \qquad (6.125)$$

where N_k^+ and N_k^- are the upper and lower noise-current sidebands, respectively, around the kth harmonic. Similarly,

$$\delta V_k = \frac{1}{2}[\Delta V_k^+\exp(j(k\omega_p + \omega_0)) + \Delta V_k^-\exp(j(k\omega_p - \omega_0))] \qquad (6.126)$$

$$\delta\omega = \Delta\Omega_k^+\exp(j\omega_0) + \Delta\Omega_k^-\exp(-j\omega_0) \qquad (6.127)$$

Substituting (6.124) through (6.127) into (6.122) and separating the terms into a matrix form gives

$$\begin{bmatrix} \mathbf{A} & \mathbf{B} \\ \mathbf{C} & 0 \end{bmatrix} \begin{bmatrix} \Delta\mathbf{V} \\ \Delta\Omega \end{bmatrix} = \begin{bmatrix} \mathbf{N} \\ 0 \end{bmatrix} \tag{6.128}$$

where **A** is an ordinary conversion matrix, as described in Section 6.1.3, and **C** accounts for the condition (6.119). Essentially, the formulation adds a phase term to the conversion-noise term.

Finally, we must determine the noise statistics. The noise component at any harmonic, k, is

$$v_k(t) = \mathrm{Re}\left\{ \sum_{k=0}^{K} [V_k^+ \exp(j\omega_s^+) + V_k^- \exp(j\omega_s^-)] \exp(j\phi_k(t)) \right\} \tag{6.129}$$

where

$$\begin{aligned} \omega_s^+ &= k\omega_p + \omega_0 \\ \omega_s^- &= k\omega_p - \omega_0 \\ \phi_k(t) &= k\int_{-\infty}^{t} \mathrm{Re}\{\Delta\Omega \exp(j\omega_0\tau)\} d\tau \end{aligned} \tag{6.130}$$

For small phase, $\tan^{-1}(\phi) \sim \phi$, so we obtain, for the phase sidebands,

$$\begin{aligned} V_k^+ &= \Delta V_k^+ + \frac{kV_{s,\,k}\Delta\Omega}{2\omega_0} \\ V_k^- &= \Delta V_k^- + \frac{kV_{s,\,k}\Delta\Omega^*}{2\omega_0} \end{aligned} \tag{6.131}$$

and $V_{s,\,k}$ is the steady-state voltage component at the kth harmonic. These can be put in matrix form:

$$\mathbf{V}_{sb} = \begin{bmatrix} \mathbf{1} & \mathbf{0} \\ \mathbf{0} & \mathrm{diag}\left(\dfrac{mV_{s,m}\Delta\Omega}{2\omega_0}\right) \end{bmatrix} \begin{bmatrix} \Delta\mathbf{V} \\ \Delta\Omega \end{bmatrix} \tag{6.132}$$

where $\mathbf{V}_{sb}$ is the vector of sideband voltages and, of course, $\Delta\Omega$ must have the proper conjugate form shown in (6.131) for lower sidebands. As before, we need the sideband voltage, $\overline{V_k^2}$. This is found in the usual manner from the correlation matrix:

$$\mathbf{C}_v = \mathbf{KMNN}^{*\mathrm{T}}\mathbf{M}^{*\mathrm{T}}\mathbf{K}^{*\mathrm{T}} \tag{6.133}$$

and all quantities of interest can be found directly from (6.133) in the same manner as with previous analyses.

It is interesting to note that the top row of (6.128) is, essentially, conversion noise analysis with an added phase term. This appears to contradict the claim in [6.8] that conversion and modulation noise are simply two different, and equivalent, views of the same phenomenon. At the same time, [6.11] contradicts both [6.8] and [6.4]. These differences point to the relative immaturity of nonlinear noise-analysis theory and illustrate the need for further research.

References

[6.1] S. A. Maas, *Nonlinear RF and Microwave Circuits*, Norwood, MA: Artech House, 2003.

[6.2] K. Kundert, J. White, and A. Sangiovanni-Vincentelli, *Steady-State Method for Simulating Analog and Microwave Circuits*, Boston: Kluwer, 1990.

[6.3] V. Rizzoli and A. Neri, "Harmonic Balance Analysis of Multitone Autonomous Nonlinear Microwave Circuits," *IEEE International Microwave Symposium Digest,* p. 107, 1991.

[6.4] E. Ngoya, J. Rousset, and D. Argollo, "Rigorous RF and Microwave Oscillator Phase Noise Calculation by Envelope Transient Techniques," *IEEE International Microwave Symposium Digest,* 2000.

[6.5] V. Rizzoli, "Optimization-Oriented Design of Free-Running and Tunable Microwave Oscillators by Fully Nonlinear CAD Techniques," *Int. J. Microwave and Millimeter-Wave Computer-Aided Engineering,* Vol. 7, p. 52, 1997.

[6.6] C.-R. Chang and M. B. Steer, "Computer-Aided Analysis of Free-Running Microwave Oscillators," *IEEE Trans. Microwave Theory Tech.*, Vol. MTT-39, p. 1735, 1991.

[6.7] E. Ngoya et al., "Steady State Analysis of Free or Forced Oscillators by Harmonic Balance and Stability Investigation of Periodic and Quasi-Periodic Regimes," *Int. J. Microwave and Millimeter-Wave Computer-Aided Engineering,* Vol. 5, p. 210, 1995.

[6.8] V. Rizzoli, F. Mastri, and D. Masotti, "A General-Purpose Harmonic-Balance Approach to the Computation of Near-Carrier Noise in Free-Running Microwave Oscillators," *IEEE International Microwave Symposium Digest,* p. 309, 1993.

[6.9] V. Rizzoli, F. Mastri, and D. Masotti, "General Noise Analysis of Nonlinear Microwave Circuits by the Piecewise Harmonic-Balance Technique," *IEEE Trans. Microwave Theory Tech.*, Vol. MTT-42, p. 807, 1994.

[6.10] S. Heinen, J. Kunisch, and I. Wolff, "A Unified Framework for Computer-Aided Noise Analysis of Linear and Nonlinear Microwave Circuits," *IEEE Trans. Microwave Theory Tech.*, Vol. MTT-39, p. 2170, 1991.

[6.11] P. Bolcato et al., "A Unified Approach of PM Noise Calculation in Large RF Multitone Autonomous Circuits," *IEEE International Microwave Symposium Digest,* 2000.

[6.12] J. M. Paillot et al., "A General Program for Steady-State, Stability, and FM Noise Analysis of Microwave Oscillators," *IEEE International Microwave Symposium Digest,* p. 1287, 1990.

[6.13] D. Scherer, "Today's Lesson—Learn About Low-Noise Design," *Microwaves*, p. 116, April 1979.

[6.14] E. Ngoya, J. Sombrin, and J. Rousset, "Simulation de Circuits et Systemes: Methodes Actuelles et Tendances," *Seminaire Antennes Actives MMIC*, Arles, p. 171, 1994.

[6.15] "Method for Simulating a Circuit," U.S. Patent No. 5588142, December 24, 1996.

[6.16] V. Rizzoli, A. Neri, and F. Masri, "A Modulation-Oriented Piecewise Harmonic-Balance Technique Suitable for Transient Analysis and Digitally Modulated Signals," *26th European Microwave Conference Digest*, p. 546, 1996.

Chapter 7

Low-Noise Amplifiers

The history of high-frequency receivers is largely the history of the search for low noise. One of the most important goals—and successes—of the MIT Radiation Laboratory, during the 1940s, was the improvement of the sensitivity of radar receivers. Throughout the 1960s and 1970s, as space communication matured, the search for low-noise diode mixers became critical. The development of low-noise microwave FET devices, and their improvement to the point of astonishingly low noise figures, was one of the major technological successes of the 1980s and 1990s.

Even though we now have extraordinarily good low-noise transistors, the problem of optimization is just as important as in the past. Indeed, there is no point in incurring the huge expense and effort of developing such devices if designers do not create circuits that realize those devices' potential. This chapter addresses that subject.

In this chapter, we are concerned exclusively with high-frequency noise. Low-frequency noise, both $1/f$ and burst noise, although present in the devices we discuss, is invariably too low in frequency to affect high-frequency amplifiers. These noise sources become important in nonlinear circuits, especially oscillators, which we examine in other chapters.

7.1 FUNDAMENTAL CONSIDERATIONS

7.1.1 Solid-State Devices

Both bipolar and FET devices are used in high-frequency amplifiers. Homojunction bipolar devices and silicon FET devices are generally suitable for use at frequencies up to a few gigahertz; some advanced technologies, especially short-gate silicon MOS devices, may be useful well into the microwave or even millimeter regions. The minimum noise figures of devices realized in III-V technologies are significantly lower than those in silicon, however. This is especially true of high-electron-mobility transistors (HEMTs) and pseudomorphic HEMTs, sometimes called pHEMTs. These offer extraordinarily low noise figures. The most advanced pHEMT technologies can operate into the high end of the millimeter-wave region. Heterojunction bipolar devices (HBTs) are also used at microwave frequencies. Their noise figures are inferior to FETs, however, so they are used primarily for low-distortion and large-signal operation.

At this writing, HEMT devices have almost completely supplanted conventional GaAs MESFETs for low-noise microwave applications. The higher electron mobility in HEMTs, combined with process optimization to reduce resistive parasitics, is largely responsible for their low noise figures. This advantage comes at a cost: HEMT devices operate at very low currents, giving them, at best, limited ability to handle large signals. They also exhibit a stronger transconductance nonlinearity, giving them higher levels of distortion than conventional MESFETs.

7.1.2 DC Bias

In both FET and bipolar devices, the minimum noise figure and other noise parameters (Section 5.2) are functions of the drain or collector current, as appropriate. As we shall see in Section 7.1.3, the noise figure is, among other effects, a function of (1) the input mismatch and (2) an interplay between the gain and output noise current. We examine these effects as they apply to FET and bipolar devices.

7.1.2.1 Bipolar Devices

In all bipolar devices, the dominant high-frequency noise source, beyond the obvious thermal noise of parasitic resistances, is shot noise in the collector current. The mean-square value of shot-noise current, from (2.53), is proportional to dc collector bias current, I_{cc}. The transconductance, g_m, at low to moderate collector currents, is proportional to I_{cc} as well, but the gain is proportional to g_m^2. Thus, at small collector currents, noise figure decreases as I_{cc} increases.

As I_{cc} increases further, the gain peaks and eventually decreases, so noise figure increases. This peak has a number of causes. In all devices, the gain cannot increase significantly beyond the point where g_m exceeds the inverse of the emitter parasitic resistance. Thermal effects and high-level injection effects (in silicon devices) also decrease the gain at high current.

As an example, Figure 7.1 shows the measured gain ($|S_{21}|^2$) and minimum noise figure of a small-signal silicon BJT at 2 GHz as a function of collector current. The gain peak occurs at 10 mA, but leveling is evident at somewhat lower current levels. The noise figure exhibits a broad minimum in the 2- to 6-mA range.

Bipolar devices, especially in silicon homojunction technologies, have relatively high base resistances compared to the gate resistances of microwave FETs. This makes the base resistance a significant noise source. The greater input resistances, along with their lower current gain-bandwidth products, ω_t (Section 7.1.3.2), are responsible for their higher noise figures.

7.1.2.2 FET Devices

As with bipolar devices, the noise figure of a FET is established by an interplay between gain, drain noise current, and thermal noise of the parasitic resistances, especially the gate and source resistances. In low-frequency MOSFETs and JFETs, the channel noise arises largely from thermal noise in the undepleted channel. In MESFETs, HEMTs, and short-channel silicon devices, however, high-field diffusion noise increases the channel noise beyond the thermal component.

The transconductance of most FETs is a much weaker function of bias current than in bipolars, but, especially in high-frequency devices, the nonthermal channel noise is a relatively strong function. As a result, the optimum noise figure of most FETs occurs at a low current, typically 15% to at

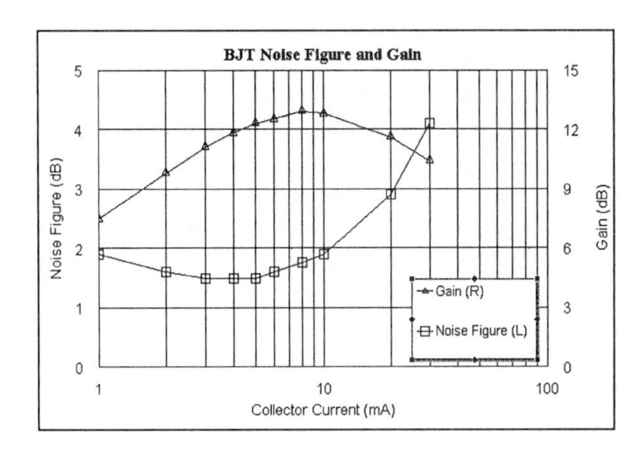

Figure 7.1 Measured noise figure and gain of a silicon bipolar transistor chip at 2 GHz.

most 25% of the maximum drain current. In microwave FETs, this is approximately 1 to 2 dB below the current that provides maximum gain and well below the level that provides the lowest distortion. Optimizing the noise figure of a FET amplifier frequently involves a painful trade-off between gain, distortion, and noise.

The gate and source parasitic resistances of microwave FETs are much lower than the base and emitter resistances of microwave BJTs. This, and the substantially higher current gain-bandwidth product, in spite of greater nonthermal drain noise, is the main reason for the superior performance of microwave FET devices.

7.1.3 Low-Noise Matching

7.1.3.1 Fundamental Considerations

In this section we address the problem of optimizing an amplifier's matching networks. We assume that the transistor can be treated as a linear two-port, and that its noise parameters and scattering parameters are available to the designer.

In Section 5.2 we made the point that the noise figure of a two-port depends on its noise parameters, $F_{\min}$, R_n, and $Y_{s,\,\mathrm{opt}}$, and the source admittance, Y_s. The most common form of the expression relating noise figure, F, to source admittance is

$$F = F_{\min} + \frac{R_n}{G_s}\left|Y_s - Y_{s,\,\mathrm{opt}}\right|^2 \tag{7.1}$$

where $F_{\min}$ is the minimum noise figure, R_n is the noise resistance, $Y_{s,\,\mathrm{opt}}$ is the source admittance that provides minimum noise figure, and $G_s = \mathrm{Re}\{Y_s\}$. The noise parameters vary with both dc bias and frequency. Beyond device selection, the noise parameters are largely out of the designer's control, so optimization of low-noise amplifiers largely involves optimizing the dc bias and source admittance. This task is straightforward in the case of narrowband amplifiers, but somewhat less obvious for broadband amplifiers.

Noise figure is independent of lossless output matching. Real matching networks invariably have loss. Input loss must, of course, be minimized; often it can be kept low enough to have a negligible effect on the amplifier's noise figure. The effect of output loss on noise figure is usually negligible, unless resistive loading or some other lossy technique (e.g., to enhance stability) is used in the output network. Then the noise introduced by the loading may be significant, although still much less than the same loss located at the input. Many amplifiers can be stabilized by adding resistors in either the input or output matching circuits; clearly, in low-noise amplifiers, they should be used only in the output, if at all.

Feedback can affect the noise parameters of a two-port (Section 5.2). Feedback can sometimes improve the input VSWR when the amplifier is noise-matched; one common technique is the use of inductance in series with the source terminal. Source inductance increases the real part of the transistor's input impedance, while affecting the minimum noise figure only slightly. Source inductive feedback is used almost exclusively in FET amplifiers; it is generally not practical or necessary in bipolar amplifiers.

7.1.3.2 Optimum Source Admittance

Realizing a source impedance as close as possible to $Y_{s,\,\mathrm{opt}}$ is the most important aspect of low-noise amplifier design. Thus, we need to know $Y_{s,\,\mathrm{opt}}$. $Y_{s,\mathrm{opt}}$ can be measured or, with a noise model of the device, calculated. Synthesizing the optimum $Y_{s,\,\mathrm{opt}}$ over a prescribed bandwidth, with minimum circuit loss, is then an exercise in matching-circuit design.

Intuitively, one might expect $Y_{s,\mathrm{opt}}$ to be a conjugate match to the device's input impedance. It may be surprising to discover that, in general, it is not; in fact, minimizing noise figure usually requires a significant input mismatch. We can demonstrate the reason for this situation through an heuristic examination of a FET's noise equivalent circuit.

To demonstrate the effect of input matching, we examine the simple FET amplifier circuit of Figure 7.2. The circuit does not include some important parasitics, notably the source-terminal resistance and feedback capacitance; these would complicate the analysis and its interpretation, but would not change the fundamental results. Noise is modeled in a manner similar to that of Pospieszalski (Section 4.4.3). The circuit contains two noise sources: a thermal noise source at the input associated with the resistance R_i, v_{ni}, and a nonthermal noise source between the drain and source, representing channel noise, i_{nd}. We assume that the input reactance is matched, so $\omega L_s = 1 / C_{gs}\omega$; this is clearly optimum, a consequence of removing all of the usual feedback elements. If we had included a feedback capacitance and source-lead inductance and resistance, a different reactance might be needed. We also assume the output to be conjugate matched, so $R_L = R_{ds}$ and $\omega L_L = 1 / C_{ds}\omega$. This latter assumption is not strictly necessary, but it simplifies the analysis and interpretation of the results.

We begin by setting $i_{nd} = 0$ to examine the effect of input matching in the absence of channel noise. Since the circuit to the right of the input loop is noiseless, we need not include it in the calculations, and we need only

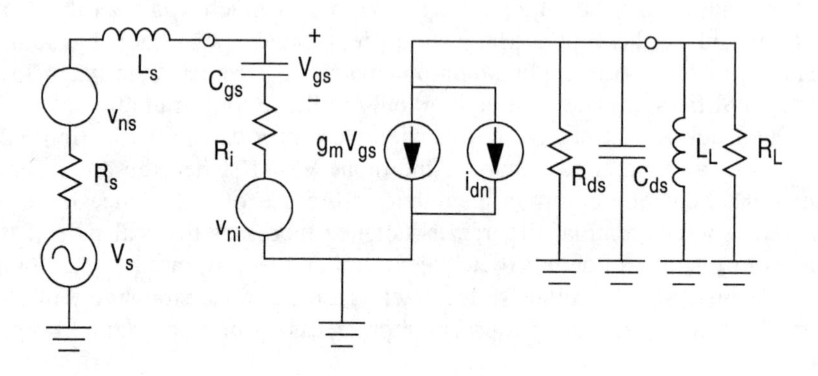

Figure 7.2 Simplified FET equivalent circuit used in the noise analysis of Section 7.1.3.2.

consider the voltage across C_{gs} as a function of the noise voltages. The noise figure [(3.9) and (5.22)] is

$$F = \frac{\overline{|i_{L,\,\text{tot}}|^2}}{\overline{|i_{L,\,s}|^2}} = \frac{\overline{|v_{c,\,\text{tot}}|^2}}{\overline{|v_{c,\,s}|^2}} \tag{7.2}$$

where $i_{L,\,\text{tot}}$ is the total output current in the load and $i_{L,\,\text{s}}$ is that portion of the current engendered by the source noise. $v_{c,\,\text{tot}}$ and $v_{c,\,s}$ are analogous voltages at C_{gs}. Then

$$v_{c,\,\text{tot}} = \frac{v_{ni} + v_{ns}}{(R_i + R_s)C_{gs}\,j\omega} \tag{7.3}$$

Since v_{ni} and v_{ns} are uncorrelated, they combine in a mean-square sense:

$$\overline{|v_{c,\,\text{tot}}|^2} = \frac{\overline{|v_{ni}|^2} + \overline{|v_{ns}|^2}}{(R_i + R_s)^2 C_{gs}^2\,\omega^2} \tag{7.4}$$

Similarly,

$$\overline{|v_{c,s}|^2} = \frac{\overline{|v_{ns}|^2}}{(R_i + R_s)^2 C_{gs}^2 \omega^2} \tag{7.5}$$

and from (7.2),

$$F = 1 + \frac{\overline{|v_{ni}|^2}}{\overline{|v_{ns}|^2}} = 1 + \frac{R_i}{R_s} \tag{7.6}$$

This result tells us that, under the stated assumptions, we can achieve an arbitrarily low noise figure by increasing R_s. In fact, a conjugate match ($R_i = R_s$) produces an uninspiring 3-dB noise figure. This intuitively unsatisfying conclusion, that an input mismatch can produce an arbitrarily low noise figure, occurs only because we have assumed away the output noise. In any case, it is something of a Pyrrhic victory; the mismatch $R_s \gg R_i$ eventually reduces the amplifier's transducer gain to impractically low values. Furthermore, as the gain decreases, the signal level in the drain circuit likewise decreases, so the effect of any noise in the drain circuit becomes progressively greater. We shall examine this effect in detail shortly.

Still, this exercise shows that mismatching the input of an amplifier, to some degree, is necessary for optimizing the noise figure. It is not difficult to see why this is so. The input signal voltage, V_s, is

$$V_s = \sqrt{8 P_{av} R_s} \tag{7.7}$$

so by increasing the source impedance, with a constant available power P_{av}, we effectively increase the signal voltage relative to v_{ni}. This increases the signal-to-noise ratio, and that increase is reflected in a decrease in noise figure. In fact, many kinds of low-frequency FET amplifiers, where the maximum available gain is high and C_{gs} is small, achieve surprisingly low noise figures in this manner.

The need for a high source impedance is a consequence of the series structure of the input circuit. If our input model were a shunt conductance with a current source, a low source impedance would optimize the noise figure.

We can derive a more realistic expression for the noise figure by including the drain noise source, i_{nd}. We now have

$$\overline{|i_{L,\,\mathrm{tot}}|^2} = \frac{1}{4}\left(g_m^2 \frac{\overline{|v_{ni}|^2} + \overline{|v_{ns}|^2}}{(R_i + R_s)^2 C_{gs}^2\,\omega^2} + \overline{|i_{nd}|^2}\right) \tag{7.8}$$

$$|i_{L,\,s}|^2 = \frac{1}{4}\left(g_m^2 \frac{\overline{|v_{ns}|^2}}{(R_i + R_s)^2 C_{gs}^2\,\omega^2}\right) \tag{7.9}$$

We obtain, from (7.2),

$$F = 1 + \frac{R_i}{R_s} + \frac{(R_i + R_s)^2 C_{gs}^2 \omega^2 G_{nd}}{g_m^2 R_s} \tag{7.10}$$

where G_{nd} is the noise conductance associated with i_{nd} (5.18). It is illustrative to convert (7.10) into the form,

$$F = 1 + \frac{R_i}{R_s} + \left(\frac{\omega^2}{\omega_t^2}\right)\frac{(R_i + R_s)^2 G_{nd}}{R_s} \tag{7.11}$$

where

$$\omega_t = \frac{g_m}{C_{gs}} \tag{7.12}$$

ω_t is the current gain-bandwidth product of the device. This quantity, expressed as the temporal frequency $f_t = \omega_t/2\pi$, is a commonly used figure of merit.

Even though it applies to a simplified equivalent circuit, (7.11) provides a valid, intuitive sense of how a FET's noise figure depends on source impedance and frequency. As R_s increases, the term R_i / R_s decreases and rapidly becomes negligible, but the more complex term increases asymptotically in proportion to R_s. At low frequencies, where $\omega \ll \omega_t$, it is possible to minimize the noise figure by the use of a large value of R_s, but at high frequencies, the significance of the latter term limits the allowable values.

The optimum value of R_s can be determined by differentiating (7.11). This exercise results in an expression that is probably too complex to be intuitively useful. Instead, we calculate the noise figure as a function of fre-

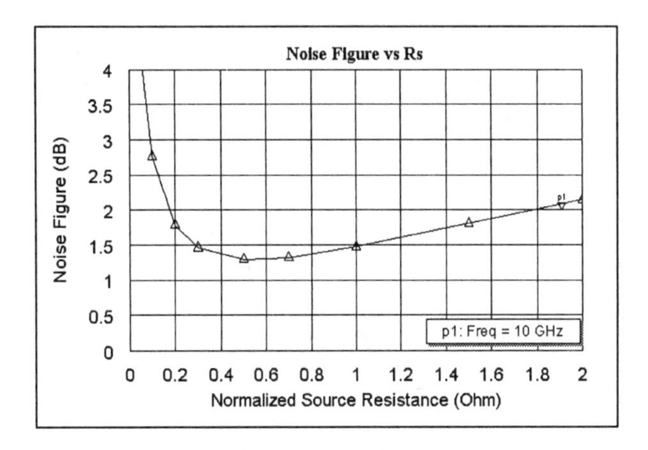

Figure 7.3 Noise figure of the circuit in Figure 7.2 at 10 GHz as a function of R_s, normalized to 50Ω. The circuit parameters are $R_i = 4\Omega$, $C_{gs} = 0.25$ pF, $g_m = 100$ mS, and $i_{nd} = 60$ pA / Hz$^{0.5}$.

quency and R_s numerically; the result is shown in Figure 7.3. We also plot $Y_{s,\,opt}$ as a function of frequency. The result, which looks much like many FETs' measured data, is shown in Figure 7.4(a).

F_{min} and R_n are shown in Figure 7.4(b). The plot of F_{min} may be different from what is normally found on transistor data sheets. Data sheets usually show a constant value of F_{min} at low frequencies. In fact, as the foregoing derivation showed, a microwave FET is virtually an ideal voltage-controlled current source at low frequencies, so its minimum noise figure theoretically approaches 0 dB. Unfortunately, $R_s \rightarrow \infty$ achieves this. Transistor data sheets instead show measured data, which is limited to practical values of R_s.

Figure 7.4(b) gives some insight as to the difficulty in optimizing the input match of a FET. The noise resistance, R_n, is relatively flat with frequency, approximately 27Ω. The theoretical minimum noise resistance, given by (5.29), varies smoothly from approximately 13 to 18 ohms from the low end of the frequency range to the upper end. This relatively high noise resistance makes FETs difficult to optimize over a broad bandwidth. Because of the inevitable difficulties in broadband matching, broadband FET amplifiers are usually matched best at the high end of the passband, where F_{min} is highest, and suffer greater noise-figure degradation at the low end, where F_{min} is lower and imperfection is more tolerable.

This analysis also provides some insight into the dependence of noise figure on bias. Up to a point, the FET's transconductance—and, thus, its ω_t—increases with drain current. This is especially pronounced in HEMT devices, which exhibit a distinct peak in transconductance with drain current. Unfortunately, G_{nd} also increases rapidly with drain bias current.

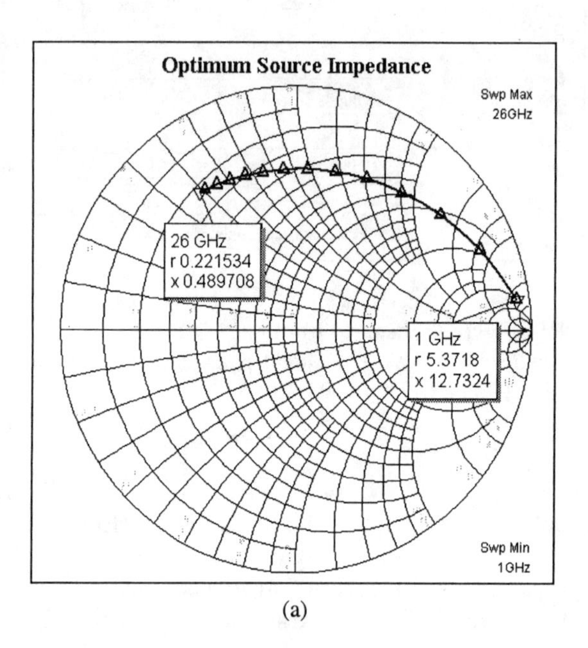

(a)

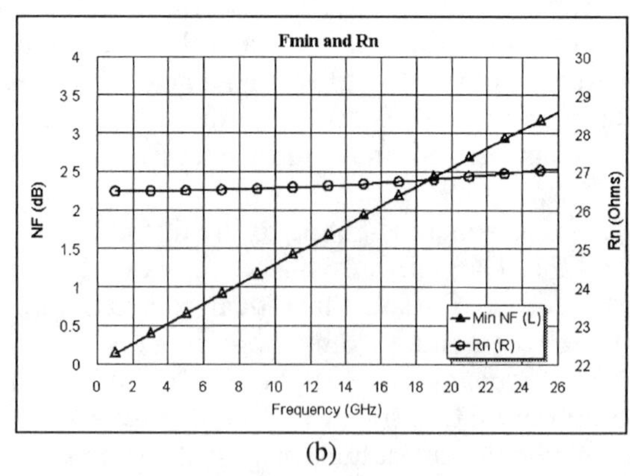

(b)

Figure 7.4 (a) Optimum source impedance as a function of frequency for the circuit in Figure 7.2, 1 to 26 GHz; (b) F_{min} and R_n. The circuit parameters are $R_i = 4\Omega$, $C_{gs} = 0.25$ pF, $g_m = 100$ mS, and $i_{nd} = 60$ pA / Hz$^{0.5}$.

These effects result in a relatively low optimum bias current, usually 15% to 25% of the peak drain current. This is well below the current required for low distortion, and results in gain 1 to 2 dB lower than is achievable at higher bias current.

7.1.3.3 Induced Gate Noise

Induced gate noise exists in high-frequency FETs, but its significance is still unclear. The Pucel model (Section 4.4.4) was developed with gate noise a clear consideration, but the Pospieszalski model (Section 4.4.3) does not include it. BSIM4, however, (Section 4.3.4) includes an induced-noise component.

The effect of gate noise is to include a component of noise in the gate that is correlated with the drain noise. If such noise is indeed significant, it complicates the above analysis, which now must account for the correlation. This modifies the value of the optimum source impedance somewhat, but, because the correlation is relatively small, the general conclusions of that analysis are still valid.

7.1.4 Input Losses

We noted in Chapter 3 that, in a cascade of two-ports, the first stage dominates in establishing the noise figure. When the first stage is lossy, its effect is especially deleterious. This point is evident immediately from Friis' formula for the noise figure of cascaded stages, (3.12). Although this formula was intended for matched stages, it is still valid for mismatched stages if the gain is defined as the available gain of each two-port [7.1]. The available gain, G_a, is

$$G_a = \frac{P_{\mathrm{av},\,o}}{P_{\mathrm{av},\,s}} \tag{7.13}$$

where $P_{\mathrm{av},\,s}$ is the power available from the source and $P_{\mathrm{av},\,o}$ is the power available from the output port of the two-port.

Losses can arise from power dissipation in imperfect components of the matching network or from radiation. Mismatch losses (such as an imperfect connector interface) do not introduce noise, but they can increase the noise figure by changing the source admittance presented to the device. Radiation losses are especially troublesome to deal with. Any circuit that radiates can also act as an antenna, receiving nonthermal noise and interference of various kinds. Placing the circuit in a metal housing may reduce such spurious reception, but the metal cover and sidewalls can also change the characteristic impedances and phase velocities of microstrip conductors, thus changing the source impedance presented to the device. A housing can also increase input-to-output coupling, which can degrade stability and change the values of the device's noise parameters. Although a metal housing eliminates radiation from matching circuits, the design of the matching circuits should account for its effects.

7.1.4.1 Effect of Loss on Matching

In the design of a low-noise amplifier, we attempt to create a lossless input matching circuit. Then, the admittance of the source (i.e., the standard admittance Z_0) is transformed to some impedance $R_s + j X_s$ at its output terminals. The available power from that transformed source impedance is the same as that from the standard source.

Real matching circuits, of course, always have some loss. If the loss is small, the transformed impedance is not changed significantly, and the only effect is the increase in noise figure given by Friis' formula, with appropriate consideration for the points made in Section 3.1.5. If losses are greater, however, the source impedance $R_s + j X_s$ is also somewhat perturbed from the lossless value. In that case, generalizations are difficult, and a complete analysis of the input circuit is necessary to predict the amplifier's noise figure.

7.1.4.2 Lumped-Element Circuits

In many types of amplifiers, lumped-element matching, or integration of lumped elements with distributed circuits, may be practical. Lumped-element circuits are helpful in minimizing size, especially in integrated circuits and at low frequencies.

Although they are theoretically reactive elements, all types of lumped capacitors and inductors have inherent losses. The losses in wirewound inductors come primarily from the resistance of the wire, increased at high frequencies by skin effect. Large inductors can radiate, so they usually must be enclosed in some way. Losses in chip capacitors result from both the resistivity of their plates and loss in their dielectrics. Although dielectric loss usually dominates, both can be significant, depending upon frequency and the structure of the component.

In any matching circuit, certain components carry higher currents than others. The high-current components are usually at the low-impedance end of the matching circuit. These components should have the highest Q possible. Using a pair of parallel-connected chip components in high-current parts of the circuit, instead of a single component, often reduces losses, because the series resistance of chip components often is relatively constant with their values. This is especially true of capacitors (less so of wire-wound inductors), so using two half-value capacitors in parallel decreases their combined series resistance to approximately half that of a single capacitor.

Planar lumped elements realized in integrated-circuit processes are notoriously lossy. In planar spiral inductors, metal losses are relatively high because defects, such as edge roughness in the conductors, are relatively large compared to the metal's dimensions, and substrates (especially silicon) are lossier. Capacitors can be lossy, compared to discrete chips, in part because of greater dielectric losses.

Both inductors and capacitors have reactive parasitics as well as resistive ones. Parasitics are almost always significant in high-frequency, lumped-element matching circuits, so they must be included in the circuit design.

7.1.4.3 Microstrip Losses

Microstrip transmission lines are used for circuit-board interconnections as well as for matching-circuit elements. Losses in microstrip arise in the metal's resistivity and in dielectric losses in the substrate. Modern substrate materials used in hybrid circuits have low loss, so metal losses invariably dominate. The same is largely true of microstrips in GaAs and InP ICs; in silicon ICs, however, the dielectric is notoriously lossy. Microstrip lines can also radiate, especially from such discontinuities as large steps in width and from the ends of open-circuit stubs.

A number of phenomena affect the losses in microstrip transmission lines. The first is simply the resistance of the line. A wider line has lower resistance, but also a lower characteristic impedance, so, for a given power, current is greater. In spite of the increased current, however, losses in practical microstrip media generally increase with decreasing line width. Thin substrates and high dielectric constants (ε_r) result in relatively narrow lines for a given characteristic impedance. Conversely, thick substrates and low ε_r allow for wide, low-loss lines; composite materials, having ε_r in the range of 2 to 4, and fused silica, having $\varepsilon_r = 3.8$, are often preferred for low-noise amplifiers. Radiation increases, however, with thick substrates at high frequencies, and generation of higher-order modes in the lines and discontinuities is also possible.

Use of a thick metallization, up to a thickness of three skin depths, also helps to minimize loss. Because currents in microstrip lines are concentrated at the edges and undersides of the strip, edge and substrate roughness can increase the transmission-line loss. Chemical etching of the substrate can cause significant edge roughness; depositing the metal by sputtering or electroplating, or etching by ion milling, results in smoother edges. Substrate roughness can be avoided through the use of polished ceramic or inherently smooth materials, such as fused silica or sapphire.

The problem of roughness is especially severe in edge-coupled transmission lines, where edge currents are especially high and strips are often narrow. This problem arises frequently in quadrature-coupled amplifiers, where multistrip Lange couplers are used. When realized on 635-μm alumina substrates, the strip widths and spacings of such couplers are on the order of 60 μm. Depending on frequency, midband excess losses of 0.5 dB or more in the couplers alone are often observed.

The current densities in microstrip matching circuits can be surprisingly nonuniform. Certain parts of a matching circuit may have locally high current densities, while the currents in other parts are very low. A modest reduction in loss can be achieved by designing the matching circuit to minimize such hot spots; suspect areas are the inside corners of T junctions and angular bends. These often can be smoothed to improve the current uniformity. A planar electromagnetic simulator can be helpful in finding such problems.

7.1.4.4 Other Transmission Media

Other transmission media, for example, stripline or suspended substrate stripline, have occasionally been used in low-noise amplifiers. These media often have the advantage of lower loss than microstrip, but they usually are less practical. For this reason, these media have been used only occasionally, mainly for such special-purpose applications as space hardware or radio astronomy.

One structure, occasionally used for high-frequency amplifiers, is microstrip in a channel, in which a microstrip line on a thick, low-ε_r substrate is mounted in a U-shaped channel in a metal housing. This has many of the characteristics of microstrip and stripline, while allowing easy access to the strip for tuning and mounting of chip components.

The high input VSWR of a low-noise amplifier can increase the losses in the coaxial line or waveguide at the input of the amplifier. If the line is longer than one-half wavelength, the high-current regions in its standing-wave pattern have disproportionately high loss. Whatever the input VSWR, coaxial lines have relatively high loss, compared to interconnection media such as waveguide, so their use at the amplifier input must be minimized.

7.1.5 Extraneous Noise Sources

Noise can be inadvertently applied to an amplifier from other sources. One of the most common is the bias circuitry.

Frequently, dc bias (especially gate bias in a FET amplifier) is applied through a large-value resistor. This resistor has a number of beneficial properties: it can improve stability, improve input match, and limit the increase in gate current when a strong signal is applied to the amplifier. The resistor's value is selected to have minimal effect on the gain and input impedance.

Depending on the FET and the resistor's value, the resistor's noise may increase the amplifier's noise figure. If the resistor's value is much greater than the input impedance at the point where it is connected, the noise current injected into the circuit is relatively small. This case is illustrated in Figure 7.5. Z_p is the impedance measured at the point where the resistor is connected [the parallel combination of Z_d and Z_s in Figure 7.5(a)]. We assume that $R_{bb} \gg |Z_p|$; if it were not, the increased input loss would have an obvious and disastrous effect on the noise figure. Since R_{bb} then has negligible effect on matching, we can treat the device noise as a noise temperature, modeled by a noise source associated with its source impedance. This source is v_{ns} in the figure. The FET itself is then noiseless.

A simple analysis shows the mean-square noise voltage applied to the input of the device to be

$$\overline{|v_n|^2} = 4KT_p \frac{|Z_p|^2}{R_{bb}} \tag{7.14}$$

where T_p is the physical temperature of the resistor and R_{bb} is its resistance. In deriving (7.14), we observe that, although the open-circuit noise voltage of the resistor increases with R_{bb}, the voltage at the device input decreases as $1/R_{bb}$. To find the increase in noise temperature caused by R_{bb}, we replace the FET gate in Figure 7.5(b) with a short circuit (since the FET, in the equivalent circuit, is noiseless) and calculate the short-circuit output currents. The result is

$$\Delta T_n = T_p \frac{|Z_s|^2}{R_{bb}\,\mathrm{Re}\{Z_s\}} \tag{7.15}$$

showing that the increase in noise temperature is negligible when $R_{bb} \gg |Z_s|^2 / \mathrm{Re}\{Z_s\}$. Z_s is usually well known, as it is an approximation of the device's optimum source impedance.

Another potential source of noise is an active bias circuit. Active circuits can generate fairly high levels of noise, which can be surprisingly broadband. It is especially important to decouple them well from the amplifier's input.

Any type of stray pickup from environmental noise sources can increase the noise temperature of a very low-noise receiver (Section 3.3.4.5). The noise can enter the amplifier through an inadequately shielded housing

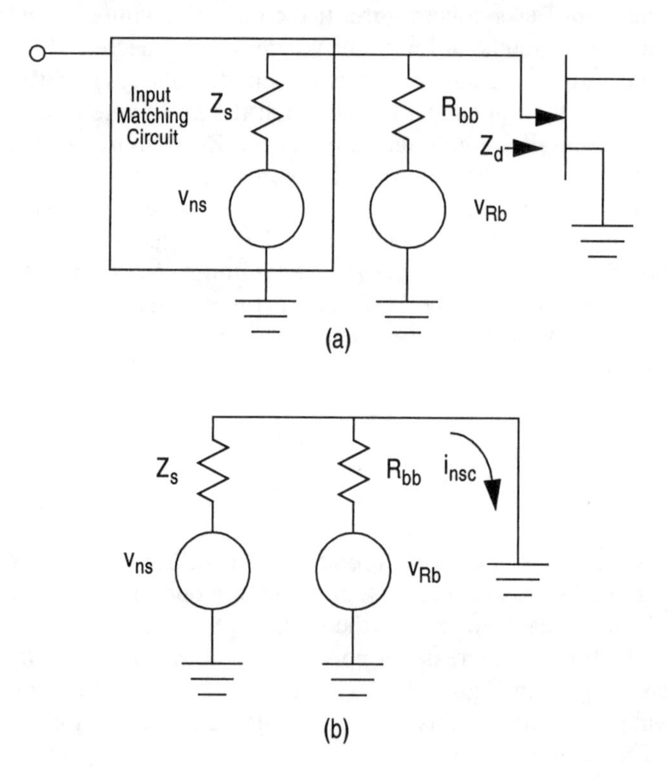

Figure 7.5 (a) Noise equivalent circuit of the input of an amplifier with a resistor used for bias insertion. The noise temperature of the amplifier is modeled by a noise source in series with the transformed source impedance; $\overline{|v_{ns}|^2} = 4KT_n\Delta f \mathrm{Re}\{Z_s\}$. (b) Equivalent circuit used to determine the increase in noise temperature caused by R_{bb}.

or via inadequately bypassed dc leads. Such receivers require great care in the design of their housings and dc circuitry.

7.1.6 The Input VSWR Problem

The need for a mismatched input creates a difficulty in the use of a low-noise amplifier in an RF or microwave system, where matched interfaces are often needed. The input mismatch can introduce gain ripple in the system's passband and increase the loss in an interconnecting cable or waveguide, thus increasing the system's noise figure. Three common methods for improving the input VSWR are feedback, an input isolator, and quadrature-coupled amplifiers.

Some of these methods affect both the input and output VSWR. In general, a high output VSWR is less of a problem than a high input VSWR, because the effect of additional losses introduced by isolators or quadrature hybrids, used to ameliorate the situation, are not as severe. We shall see, in Section 7.2.2, that the need for broad bandwidth inevitably results in high output VSWR over part of the band, and thus some method of VSWR improvement is needed at the output as well as the input.

7.1.6.1 Feedback

Reactive feedback does not decrease the noise figure to any practical degree. A classic paper by Haus and Adler [7.2] showed generally that the noise *measure* of a transistor remains constant, even when F_{min} decreases.[1] Feedback can change $Y_{s,\,opt}$ to a more desirable value, however, to facilitate the design of the input matching circuit.

In particular, inductive feedback in the source terminal of a FET increases the real part of the input impedance, making it possible to match the device simultaneously for both noise and VSWR. Such matching is, unfortunately, relatively narrowband and can affect stability. The synthesized real component of input impedance is noiseless.

A simple analysis of a FET input circuit with source-lead inductance shows that the input impedance is increased by

---

1. This piece of conventional wisdom, while generally correct in practical applications, may not be strictly true in theory. Series inductive feedback can sometimes reduce F_{min} while making the device conditionally stable. The maximum gain is then theoretically unlimited, so the noise measure equals F_{min}. When this occurs, most other aspects of the performance are poor (e.g., R_n is large and $|S_{12}|$ is great), so achieving improved noise figure, in practice, may not be possible.

$$\Delta Z_{\text{in}} = g_m \frac{L}{C_{gs}} + Lj\omega \tag{7.16}$$

where ΔZ_{in} is the increase in impedance and L is the source inductance. Although the increased $\text{Re}\{Z_{\text{in}}\}$ is welcome, it is often accompanied by increased R_n, S_{12}, and S_{22}, all of which are undesirable.

7.1.6.2 Isolators

Isolators represent a brute-force approach to the improvement of input VSWR. Isolators that have coaxial connectors are usually realized in stripline; waveguide isolators use ferrite loaded waveguide junctions. Isolators are large, heavy, costly, and introduce at least a few tenths of one decibel of additional input loss. An isolator is needed on each port that requires VSWR improvement.

Microstrip isolators, which can be mounted inside the amplifier package, are also available. These are much smaller than stripline or waveguide isolators but have greater loss. Because of the difficulty of making connections to them, they usually do not provide as good VSWR as the other types.

7.1.6.3 Quadrature-Coupled Amplifiers

One of the most common configurations for microwave amplifiers of all types is shown in Figure 7.6. In the figure, a pair of ideally identical amplifiers is coupled by quadrature hybrids at both the input and output. The hybrids' isolated ports are terminated. This configuration, first proposed by Englebrecht and Kurokawa [7.3], has a number of useful characteristics. One of the most important is that the input VSWR is ideally 1.0, regardless of the input reflection coefficients of the individual amplifiers, as long as they are identical and the hybrids are ideal. Even with the limitations that reality places upon such components, however, low input VSWR can still be achieved over a broad bandwidth.

It is easy to see why this occurs by considering the input hybrid in Figure 7.6. An ideal quadrature hybrid has the scattering matrix,

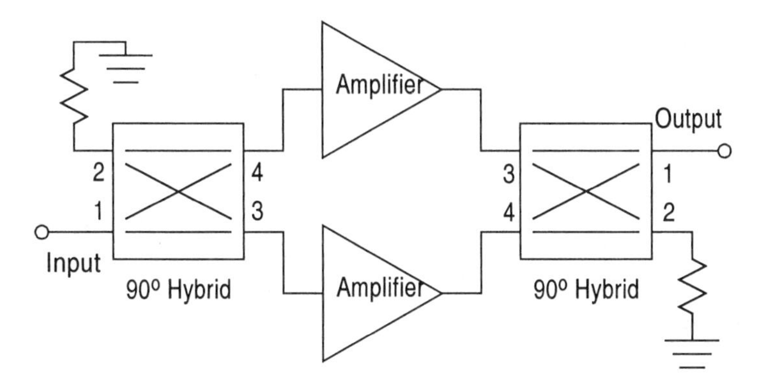

Figure 7.6 Quadrature-coupled amplifiers. The port numbering of the hybrids is for reference in the text.

$$\mathbf{S}_{90} = \frac{1}{\sqrt{2}}\begin{bmatrix} 0 & 0 & -j & 1 \\ 0 & 0 & 1 & -j \\ -j & 1 & 0 & 0 \\ 1 & -j & 0 & 0 \end{bmatrix} \tag{7.17}$$

showing that the paths through the hybrid have 3-dB loss and either 0- or 90-degree phase shift. (The phase shift through real hybrids is much greater; the difference between in phase at the output ports is the important quantity; absolute phase delay is not.) Port 1 is the input and ports 3 and 4 are terminated with the reflection coefficient of the amplifiers, Γ, assumed to be equal. The terminations constrain the a and b waves at ports 3 and 4 to be

$$a_3 = \Gamma b_3 \tag{7.18}$$

and

$$a_4 = \Gamma b_4 \tag{7.19}$$

Substituting (7.18) and (7.19) into (7.17) gives

$$\begin{bmatrix} b_1 \\ b_2 \end{bmatrix} = \Gamma \begin{bmatrix} 0 & -j \\ -j & 0 \end{bmatrix} \begin{bmatrix} a_1 \\ a_2 \end{bmatrix} \tag{7.20}$$

Port 2 has an ideal termination, so $a_2 = 0$ and therefore $b_1 = 0$. All the power reflected from the individual components is dissipated in the load at port 2, and none emerges from port 1, so the input port is matched. The same applies to the output port, which is also matched. When the terminations on ports 3 and 4 are unequal and the termination on port 2 is ideal, we can derive, similarly,

$$\Gamma_{in} = 0.5(\Gamma_3 - \Gamma_4) \tag{7.21}$$

where $\Gamma_{in} = b_1/a_1$, and Γ_3, Γ_4 are the input reflection coefficients of the two amplifiers terminating ports 3 and 4, respectively. We see that, even when the port terminations are not precisely equal, the input VSWR may still be very low.

The quadrature couplers in microwave amplifiers are realized as coupled-line structures. Because it is impossible to achieve adequate coupling in microstrip between a single pair of lines, four or more strips are used and the conductors are interleaved. These so-called *Lange hybrids* also use a clever arrangement to create a crossover in the center of the coupler, placing the input and output ports in the desired locations. Quadrature-coupled amplifiers are almost always realized in microstrip on alumina substrates, as the high dielectric constant (approximately 9.6 to 10.0, depending upon manufacturing methods) provides small size and adequate coupling. See [7.4] for further information on such couplers.

An ideal coupled-line hybrid's port VSWR, phase balance, and isolation are theoretically perfect and frequency independent. Its amplitude balance is imperfect, however, varying as $\sin^2(\pi\omega / 2\omega_0)$, where ω_0 is its center frequency. This imperfection is not as important as it may seem at first, because, if the loss from the input to one port of the hybrid is c ($c < 1$), the loss from the input to the other port must be $1 - c$. Figure 7.6 shows that a signal passing through one branch of the circuit must experience loss c through one hybrid and loss $1 - c$ in the other. The couplers largely compensate each other, giving the amplifier surprisingly flat gain, even at frequencies where the coupler's balance is poor. Furthermore, for broadband operation, the couplers can be overcoupled at bandcenter to minimize the worst-case imbalance at the band edges.

A great concern, for our purposes, is the noise from the termination connected to the unused port of the input coupler. This noise is applied to the inputs of the amplifiers, so at first appearance, the noise temperature must increase by the temperature of this termination. However, the noise from the resistor is correlated in the amplifiers' outputs, so it is subtracted by the output coupler. Ideally, none of the termination's noise reaches the output.

The termination noise can reach the output only if the hybrids are not perfectly balanced. The increase in output noise temperature arising from amplitude imbalance between the in-phase and quadrature-phase paths through the hybrids is

$$\Delta T_{nL} = G_{ta} T_t (2c - 1)^2 \tag{7.22}$$

where ΔT_{nL} is the increase in output noise temperature, T_t is the termination's physical temperature, G_{ta} is the amplifiers' gain (i.e., the amplifier blocks alone, not including the hybrids), and c is either hybrid's power coupling factor at the frequency of interest, which generally deviates from the ideal bandcenter value of 0.5. In (7.22), we assume the phase balance to be perfect; in fact, phase imbalance in practical hybrids is minor, so amplitude imbalance dominates. The imbalance reduces the gain of the complete amplifier to $4G_{ta}c(1 - c)$, so the increase in the input noise temperature, ΔT_n, is

$$\Delta T_n = \frac{T_t (2c - 1)^2}{4c(1 - c)} \tag{7.23}$$

Even this is a relatively minor increase in noise temperature in most cases.

The coupling loss has no effect on the noise figure of the amplifier; if the coupler is ideal, the noise figure of the coupled amplifiers is the same as that of the individual amplifiers. However, excess loss in the coupler, caused by resistive losses in the microstrip, increases the noise figure of the amplifier in the same manner as any input attenuation (Section 7.1.4.3). The excess coupler loss, combined with the loss in relatively narrow microstrips, can be significant. Thus, such amplifiers do not achieve the lowest possible noise figures.

7.1.6.4 Output Loading

It is well known that the output load of a nonunilateral two-port affects the input reflection coefficient. The input reflection coefficient, Γ_{in}, of a terminated two-port is

$$\Gamma_{\text{in}} = S_{11} + \frac{S_{21}S_{12}\Gamma_L}{1 - S_{22}\Gamma_L} \tag{7.24}$$

where Γ_L is the load reflection coefficient. It seems possible, in some cases, to adjust Γ_L so that Γ_{in} is a conjugate match to $\Gamma_{s,\text{opt}}$. Unfortunately, even when this is possible, it invariably results in a poor output match and gain that is not constant over the desired frequency range. In effect, this technique, when possible at all, simply displaces the VSWR problem from the input to the output. Although this might be an improvement in some cases, the method is suggested in the literature more frequently than used in practice.

7.1.7 Thermal Effects and Cooled Amplifiers

Heating and cooling have obvious effects on the thermal noise sources in low-noise devices. They affect nonthermal sources, and the amplifier's noise figure, in a somewhat indirect manner.

As a FET's temperature increases, its transconductance decreases, partly from increases in parasitic resistance (especially the source-lead resistance) and from a decrease in the electrons' saturation velocity. C_{gs} changes little, so ω_t decreases. G_{nd} increases as well, although part of the increase may come from increased drain current, which is necessary for maintaining adequate transconductance. From (7.11), F must increase with temperature. R_i in (7.11) is a thermal noise resistance, so it must be scaled in proportion to absolute temperature as well.

FET amplifiers (but not bipolar amplifiers) can be cooled to achieve great reductions in noise temperature. For such applications as radio astronomy, which require extremely high sensitivity, FETs are often cooled to cryogenic temperatures. For example, Pospieszalski [7.5] shows a decrease in minimum noise temperature of a factor of approximately 6 for an 8.5-GHz HEMT cooled from 297K to 12.5K. The author has observed noise-temperature reductions of a factor of approximately 3 in cooling a MESFET amplifier from room temperature to 77K.

Maintaining low noise figures of both cooled and room-temperature amplifiers requires care in thermal design and design of the bias networks.

The FET chip or package must have adequate heat sinking, even though the power dissipation at low-noise bias is rarely very great. To compensate for gain changes, bias circuits often increase dc drain current as temperature increases. This kind of compensation, if not performed carefully, can increase the noise figure significantly.

Both the noise and S parameters of a FET change markedly with decreasing temperature. Ideally, the amplifier should be designed according to the low-temperature S and noise parameters, and any manual tuning should be performed at the low temperature. Because of the practical difficulty of measuring S and noise parameters at very low temperatures, amplifiers intended for low-temperature operation are often designed according to room-temperature design data. The improvement on cooling such amplifiers is probably not as great as theoretically possible, but still significant.

7.2 AMPLIFIER OPTIMIZATION

In this section we address the problem of optimizing the noise performance of amplifiers. We focus entirely on aspects of low-noise design; we do not discuss the basics of amplifier design, as that subject is well covered in other texts.

7.2.1 Narrowband Amplifiers

7.2.1.1 Fundamental Considerations

The design of narrowband amplifiers—amplifiers having bandwidths of perhaps 10% or less—is straightforward. The input matching circuit is designed to present the optimum source admittance, $Y_{s,\,\mathrm{opt}}$, to the transistor; the output impedance, with the optimum noise admittance loading the input, is then calculated. Finally, the output circuit is designed to achieve a conjugate match. Designing these circuits at the center frequency usually provides adequate performance over the required bandwidth.

A perennial problem in the design of narrowband amplifiers is high gain outside of the desired band, especially at low frequencies. This characteristic is generally undesirable, as it leads to instability and allows interference from strong, out-of-band signals. Proper design of the matching and bias circuits can minimize this problem; for example, small series capacitors and shunt short-circuit stubs in the matching circuit, and resistive loading in the bias circuit, can reduce low-frequency gain.

7.2.1.2 Example

We now design a 9.5- to 10.5-GHz low-noise amplifier using a conventional high-frequency MESFET. The goal is to achieve a good output match and nearly optimum noise figure over the band. These requirements define the matching circuits, so it is impossible to specify the gain as well. The gain will be whatever results from these matching conditions; we expect it to be approximately 10 dB. The circuit is realized as a microwave hybrid on an alumina substrate, $\varepsilon_r = 9.8$.

The amplifier is designed in the following manner:

1. The first step is to check the stability of the device. The stability factor (K factor) is less than 1.0 at frequencies below 8 GHz, creating a potential for oscillation. We design the bias decoupling circuits to include some resistive loading, which should reduce the out-of-band gain enough to stabilize the amplifier. The bias circuits are designed to decouple those resistors at ~10 GHz, to prevent noise of the bias supplies and loading resistors from increasing the amplifier's noise figure. We find that the resistors improve the stability factor, making $K > 1$ at all frequencies. We also check the S parameters and minimum noise figure to make sure that they have not been changed by the bias circuit or loading resistors.

2. Next, we design the input matching circuit. The goal is to synthesize the source reflection coefficient for optimum noise match. A simple circuit consisting of series lines and open-circuit stubs suffices.

3. With the input matching network in place, we calculate the output reflection coefficient over the band of interest.

4. Finally, we design the output matching network. To minimize low-frequency gain, it includes a shorted stub.

This initial design should be close to optimum, but still may benefit from numerical optimization. An initial result that is far from optimum indicates that an error has been made in the above process; it should be fixed before any further numerical optimization is performed. It is always a bad practice to use an optimizer to fix errors.

The amplifier circuit is shown in Figure 7.7. Figure 7.8(a) shows its gain and output return loss, and Figure 7.8(b) shows its noise figure, the minimum noise figure of the device, and the input return loss. The noise

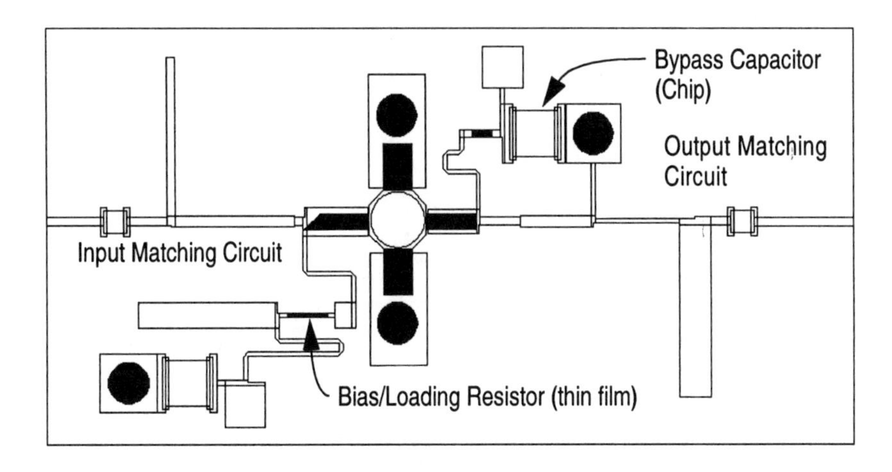

Figure 7.7 Layout drawing of the narrowband amplifier. The input and output use simple stub matching circuits, designed at the center frequency (10 GHz). Although the design bandwidth is 1.0 GHz, good performance is achieved over a 2.0 GHz bandwidth.

figure is approximately 0.5 dB worse than the device's minimum; the increase is largely caused by circuit loss.

The input VSWR is approximately 2.0 worst case, adequate for many applications. It is not difficult to see why the VSWR is unexpectedly good: we are using a mediocre device. As $\omega \to \omega_t$, the rightmost term in (7.11) becomes dominant, so it is more important to minimize it, by conjugate matching the input, than to minimize the R_i / R_s term.

7.2.2 Broadband Amplifiers

7.2.2.1 The Broadband Amplifier Problem

The need for broad bandwidth creates special problems in the design of low-noise amplifiers. The fundamental problem is the theoretical impossibility of creating a perfect input matching circuit (i.e., one that synthesizes $Y_{s,\,\mathrm{opt}}$ exactly) over a broad bandwidth. We must, instead, tolerate some degree of mismatch, somewhere within the amplifier's passband. In the input circuit, the obvious place to tolerate this mismatch is at the low end of the band, where F_{min} is lower and there is more room for error.

For the same reasons, we cannot match the output arbitrarily well over a wide bandwidth. In fact, a perfect broadband output match usually is undesirable, as it results in a sloped gain, lower at the high-frequency end of

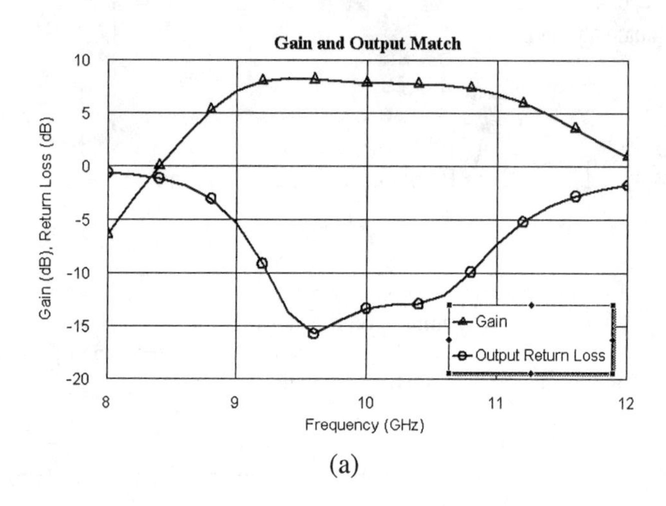

(a)

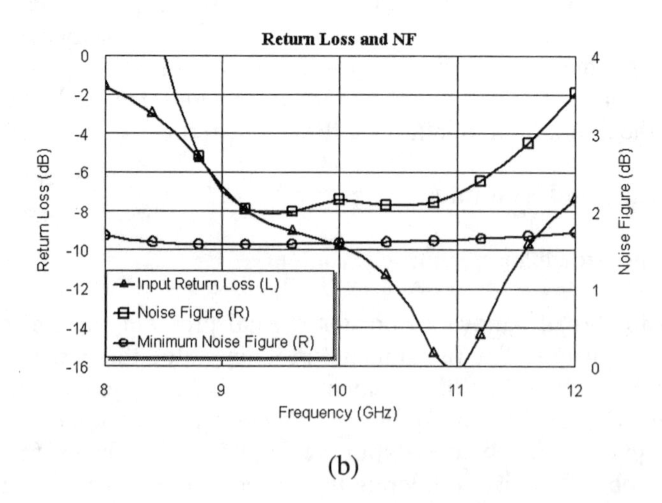

(b)

Figure 7.8 Performance of the narrow-band amplifier: (a) gain and output return loss; (b) input return loss and noise figure compared to the device's F_{min}.

the band than at the low-frequency end. Instead, we purposely mismatch the low end of the band to achieve flat gain over the desired passband.

The fundamental problem becomes one of designing the matching circuits to achieve these goals, along with the usual requirements of stability and low out-of-band gain. This can be a complicated task. One approach is to determine a set of source and load impedance loci that provide the desired performance and can be modeled simply. If such a set can be found, we can then generate the matching circuits in a straightforward manner. Fortunately a technique called *negative-image modeling* can accomplish this. We describe the method in the following section.

7.2.2.2 Negative-Image Modeling

The design of the input circuits for broadband amplifiers requires competing trade-offs between gain and noise. The fundamental problem is to determine an appropriate, realizable set of source and load reflection coefficients, over the band of interest, that optimizes all requirements of the amplifier. The method described here, first proposed by Medley and Allen in 1979 [7.6], is both elegant and practical.

The method is as follows:

1. We first create a circuit having the special negative-image source and load networks as shown in Figure 7.9(a). $-C_s$ and $-C_L$ are negative capacitances. These networks can have any structure, but, to facilitate the synthesis of the real networks, they should be as simple as possible. The source network should approximate the locus of $\Gamma_{s,\,\mathrm{opt}}$ and the output network should mirror the structure of the device's drain equivalent circuit.
2. We then optimize the circuit by means of a linear circuit-analysis program, using whatever trade-offs are appropriate. Because of the negative capacitances, the optimization is surprisingly easy.
3. When satisfactory performance has been achieved, we synthesize input- and output-matching networks using loads that are the positive versions of the negative-image networks [Figure 7.9(b)]. This is best accomplished by a circuit-synthesis program, but any other favored method can be used.
4. We replace the negative-image circuits with the matching circuits and do any necessary final optimization.

If the matching circuits synthesized in Figure 7.9(b) provided a conjugate match to their respective positive-load networks, their output impedances would be exactly equal to those of the negative-image networks. Even though a perfect conjugate match is not possible, the networks still provide a good approximation of the optimum source and load reflection coefficients, Γ_s and Γ_L, of those networks.

7.2.2.3 Example

The process of negative-image modeling is best described by a design example. We wish to design an 8- to 12-GHz low-noise amplifier. We begin by creating the negative-image model shown in Figure 7.10. This model may require a little experimentation; a series RC at the input and a parallel RC at the output usually work well, but in this case, we use a parallel RLC at the output. We optimize the circuit using these negative-image networks; the optimization emphasizes whatever aspects of the performance are im-

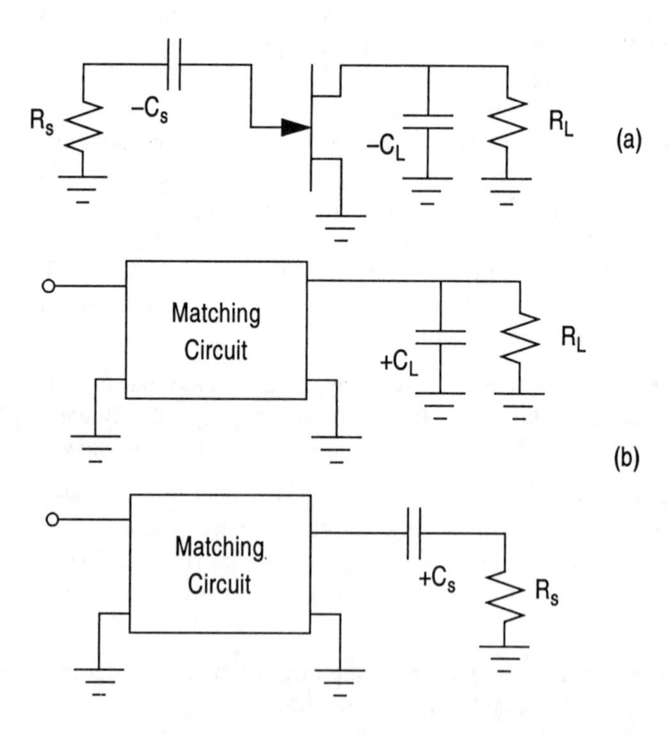

Figure 7.9 Negative-image matching: (a) a FET with negative-image networks; (b) synthesis of equivalent real matching circuits.

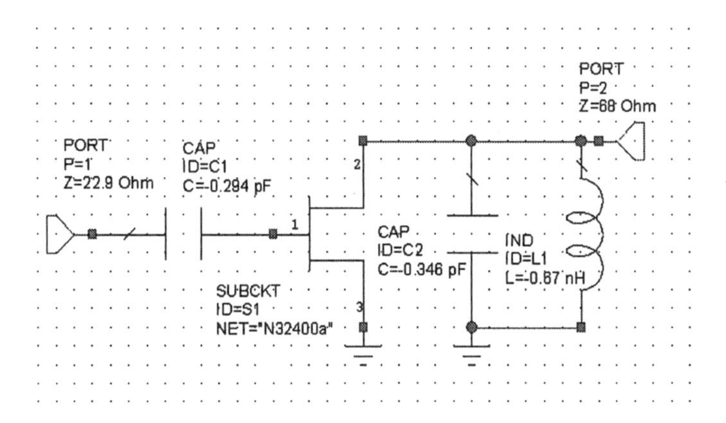

Figure 7.10 The optimized negative-image model of the amplifier.

portant for the amplifier's application. The next step is to synthesize the matching circuits, which is accomplished with the help of circuit-synthesis software. Finally, we add the synthesized circuits to the FET and do any final optimization deemed necessary.

The result, compared to the performance of the negative-image model, is shown in Figure 7.11. The midband gain of the finished amplifier is quite close to the negative-image model, but it rolls off somewhat at the high end of the band. The noise figure is largely as expected; it follows the negative image model over most of the band, diverging most strongly at the lower edge. Numerical optimization of the finished amplifier could further improve the performance.

The critical part of the design is the optimization of the circuit in Figure 7.10. The characteristics of this circuit, including stability, noise figure, gain, and all other parameters of interest are, within the limitations of matching-network synthesis, those of the resulting amplifier. The trade-offs at this stage of the design can be whatever the designer deems appropriate; usually, the negative-image model is designed with the help of numerical optimization. The goals and weights of the optimized parameters determine the performance of the finished amplifier.

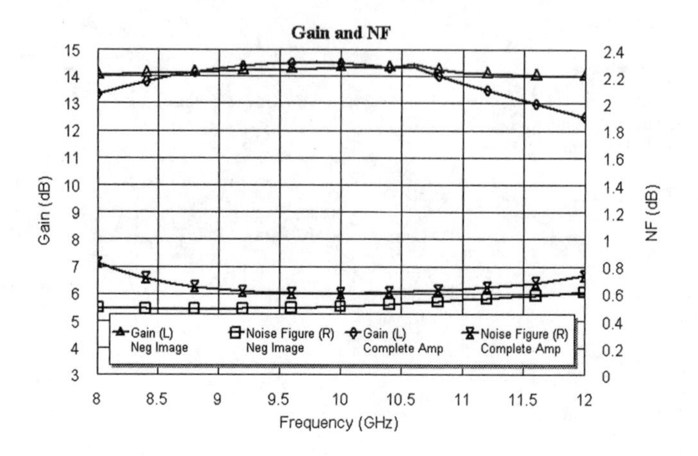

Figure 7.11 Comparison of the gain and noise figure of the finished amplifier to that of the negative-image model.

References

[7.1] E. W. Strid, "Measurement of Losses in Noise Matching Networks," *IEEE Trans. Microwave Theory Tech.*, Vol. MTT-29, p. 247, 1981.

[7.2] H. A. Haus and R. B. Adler, "Optimum Noise Performance of Linear Amplifiers," *Proc. IRE*, Vol. 46, p. 1517, 1958.

[7.3] R. S. Englebrecht and K. Kurokawa, "A Wide-Band, Low-Noise L-Band Balanced Transistor Amplifier," *Proc. IEEE*, Vol. 53, p. 237, 1965.

[7.4] R. Mongia, I. Bahl, and P. Bhartia, *RF and Microwave Coupled-Line Circuits*, Norwood, MA: Artech House, 1999.

[7.5] M. W. Pospieszalski, "Modeling of Noise Parameters of MESFETs and MODFETs and Their Frequency and Temperature Dependence," *IEEE Trans. Microwave Theory Tech.*, Vol. MTT-37, p. 1340, 1989.

[7.6] M. Medley and J. L. Allen, "Broad-Band GaAs FET Amplifier Design Using Negative-Image Device Models," *IEEE Trans. Microwave Theory Tech.*, Vol. 27, p. 784, 1979.

Chapter 8

Oscillators

With the increasing use of phase and phase-amplitude modulation in cellular telephone and radio systems, the analysis of phase noise in oscillators has become a subject of critical importance in industry. Although phase noise has always been an important subject, the development of theory and simulation technology in the past decade has allowed a quantitative treatment, and, at least in principal, the ability to minimize it.

In this chapter, we begin by treating basic oscillator theory in a more or less historical manner, showing how classical approaches lead to a more complete theory. We then examine noise theory itself, and ways to optimize these circuits.

8.1 CLASSICAL APPROACHES TO OSCILLATOR THEORY

8.1.1 Linear Stability

The conditions for stability of a linear system are simple and well known: a transfer or impedance function of the form

$$H(s) = \frac{\prod_{m}(s+z_m)}{\prod_{n}(s+p_n)} \tag{8.1}$$

must have no poles (zeros of the denominator) in the right half of the s plane. When such poles exist, the response to a small, transient excitation grows exponentially. Linear circuit theory cannot predict the degree of growth of the instability; nonlinearities in the circuit limit it. In an oscillator we want a small excitation (such as noise in the circuit or the turn-on transient) to create a growing sinusoidal response. A stable oscillation occurs when the pole is on the $j\omega$ axis; that is, the real part is zero. It is, impossible to achieve this condition exactly in a linear circuit; however, in a nonlinear circuit, it can be achieved readily.

In RF and microwave circuits, we rarely create a function of the form (8.1) or have much of an idea of the pole locations in any circuit. Thus, we are forced to examine stability, and, hence, conditions for oscillation, by other means.

8.1.2 Barkhausen Criterion

Figure 8.1 shows a model of an oscillator consisting of an amplifier stage plus feedback. The amplifier's transfer function is $A(s)$ and the feedback transfer function is $F(s)$. The transfer function of the combination is easily shown to be

$$H(s) = \frac{V_o(s)}{V_i(s)} = \frac{A(s)}{1 - A(s)F(s)} \tag{8.2}$$

On the imaginary axis, we have

$$H(j\omega) = \frac{A(j\omega)}{1 - A(j\omega)F(j\omega)} \tag{8.3}$$

which is most relevant, since the oscillation must be steady-state sinusoidal. From Section 8.1.1, oscillation occurs when

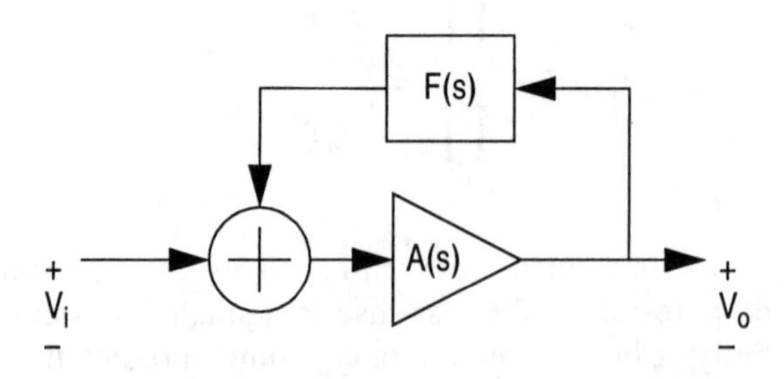

Figure 8.1 Feedback oscillator illustrating the Barkhausen criterion.

$$1 - A(j\omega)F(j\omega) = 0 \tag{8.4}$$

or

$$A(j\omega)F(j\omega) = 1 \tag{8.5}$$

That is, magnitude of the loop gain is unity when the phase is zero.

It is easy to see intuitively what happens when (8.5) is satisfied. An incremental input, having a frequency component at a frequency that satisfies (8.5), is amplified by the amplifier block. The response is fed back to the input under such conditions that it combines in phase with the input frequency component, and the output increases. The output then grows, with a characteristic time dependent on the bandwidth of the system, until the amplifier saturates. The result is a stable, oscillatory output.

The requirement for oscillation—a phase shift of 0 degrees and gain of unity—is called the *Barkhausen criterion*. Since a gain greater than unity clearly satisfies the oscillation requirement, it is often written as

$$\begin{aligned} |A(j\omega)F(j\omega)| &> 1 \\ \angle A(j\omega)F(j\omega) &= 0 \end{aligned} \tag{8.6}$$

which corresponds to poles in the right half plane. This derivation shows that the Barkhausen criterion is simply a consequence of the existence of unstable poles in the closed-loop transfer function.

Applying the Barkhausen criterion to practical circuits is usually straightforward, as long as the forward and feedback paths can be clearly identified. For example, consider the Colpitts oscillator in Figure 8.2(a). The equivalent circuit of the oscillator is shown in Figure 8.2(b); it consists of a part that provides gain (the controlled source) and one that provides feedback (the *pi* network). In this case, the oscillation condition is

$$V_i = \alpha V_c \tag{8.7}$$

where $\alpha = 1$ according to the strict Barkhausen criterion or $|\alpha| > 1$, $\angle\alpha = 0$ for the more practical situation. Introducing the loop gain, we find the oscillation condition to be

$$A(j\omega)F(j\omega) = -g_m Z_{2,1}(j\omega) = \alpha = 1 \tag{8.8}$$

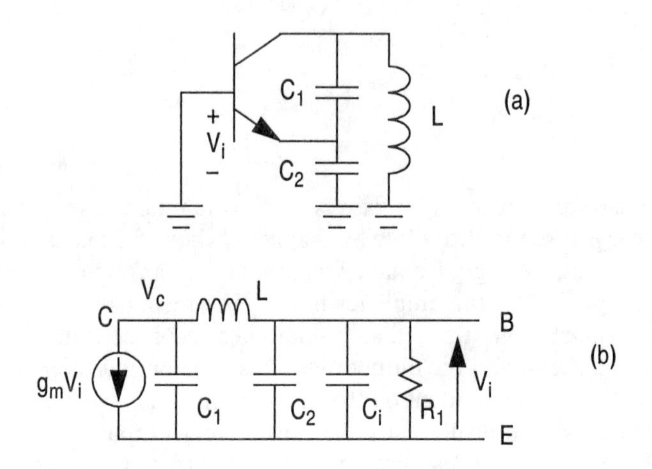

Figure 8.2 (a) A Colpitts oscillator, and (b) its equivalent circuit. C_i and R_i are the input capacitance and resistance, respectively, of the transistor.

where $Z_{2,1}$ is an impedance parameter of the RLC *pi* network.

We could also view the oscillator as follows: the nodal equations of the network are

$$\begin{bmatrix} g_m - \frac{1}{j\omega L} & \frac{1}{j\omega L} + j\omega C_1 \\ j\omega C_t + \frac{1}{R_1} + \frac{1}{j\omega L} & \frac{-1}{j\omega L} \end{bmatrix} \begin{bmatrix} V_i \\ V_c \end{bmatrix} = \begin{bmatrix} 0 \\ 0 \end{bmatrix} \tag{8.9}$$

where $C_t = C_2 + C_i$, and C_i is the input capacitance. The input resistance $R_1 = \beta / g_m$, where β is the current gain of the transistor. For a nontrivial solution, the determinant of the matrix in (8.9) must be zero. A little algebra gives

$$\beta = \frac{C_1}{C_t} \tag{8.10}$$

In practice, $\beta > C_1 / C_t$ is required. The frequency f_0 satisfying the requirement of the zero determinant is

$$f_0 = \frac{1}{2\pi}\sqrt{\frac{C_1 + C_t}{LC_1C_t}} \tag{8.11}$$

which is the frequency of oscillation. This is simply a resonance formed by the inductance and series combination of C_1 and C_t.

8.1.3 Kurokawa Theory

The classic work on oscillators is that of Kurokawa [8.1]. Kurokawa derived conditions for oscillation and stability of a negative-resistance oscillator, a two-terminal nonlinear impedance or admittance having a negative real part. This oscillator model is shown in Figure 8.3. Although this model was undoubtedly inspired by two-terminal oscillating devices, such as Gunn devices, IMPATTs, and tunnel diodes, the theory can be applied to transistor oscillators as well. In the design of a transistor oscillator, feedback creates a negative impedance or admittance at a pair of the transistor's terminals, and the design process proceeds as with a two-terminal device.

Because of the computational limitations at the time of that work, the analysis was subject to a number of simplifying assumptions. The most important was that only the fundamental-frequency component of current in Figure 8.3(a) or voltage in Figure 8.3(b) existed in the circuit. Because most oscillators use high-Q resonators, this assumption was not unreasonable.

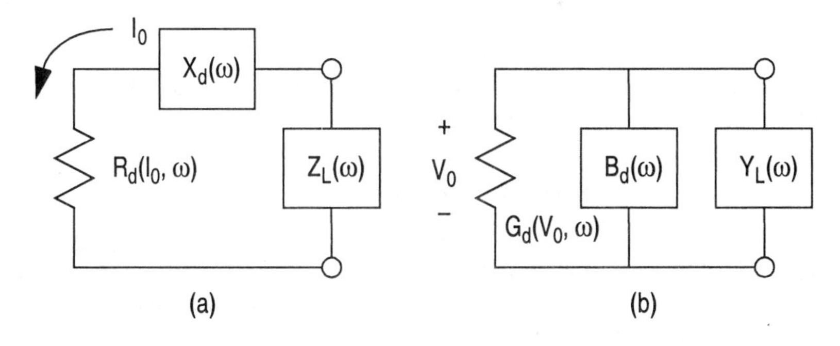

Figure 8.3 Kurokawa oscillator: (a) impedance model; (b) admittance model.

8.1.3.1 Oscillation Conditions

Kurokawa defined the admittance of the oscillating device, Y_d, as

$$Y_d(V_0, \omega) = \frac{I(\omega)}{V(\omega)} \tag{8.12}$$

where V_0 is the magnitude of the voltage across the device, assumed to be a sinusoid of frequency ω_p, and $I(\omega)$ and $V(\omega)$ are, respectively, the fundamental-frequency components of its voltage and current. This view of an impedance as the ratio of two frequency components differs from the more common idea of an incremental conductance, dI/dV, or an incremental capacitance, dQ/dV. In any nonlinear circuit, (8.12) describes a large-signal impedance, the quantity that must be conjugate matched to optimize power transfer. In any practical oscillator, the magnitude of $\mathrm{Re}\{Y\}$ decreases as V_0 increases.

He further proved that the condition for oscillation is

$$Y_d(V_0, \omega) + Y_L(\omega) = 0 \tag{8.13}$$

where $Y_L(\omega)$ is the load admittance. From (8.13), $Y_d + Y_L$ create a parallel resonance so, as long as the resonator Q is high, it is reasonable to assume that $V(\omega) = 0$ for all harmonics beyond the fundamental frequency.

It is also possible to define a series-resonant oscillator, in which the device impedance, Z_d, is

$$Z_d(I_0, \omega) = \frac{V(\omega)}{I(\omega)} \tag{8.14}$$

and I_0 is the magnitude of the current in the device. The condition for oscillation is

$$Z_d(I_0, \omega) + Z_L(\omega) = 0 \tag{8.15}$$

where Z_L is the load impedance. In this case, I_0 has only a fundamental-frequency component, and $|\mathrm{Re}\{Z_d\}|$ invariably decreases as I_0 increases. This is simply a dual case of (8.12) to (8.13).

In the design of a negative-resistance oscillator, we do not attempt to satisfy (8.13) or (8.15) exactly. Instead, we attempt to satisfy only the reactive condition at start-up (i.e., as $V_0 \to 0$ for parallel resonance or $I_0 \to 0$

for series resonance) and create "net negative resistance (or conductance)" in the circuit at start-up. The resistance or conductance then decreases as the oscillation builds, and eventually the oscillation stabilizes at the value satisfying (8.13) or (8.15). As the oscillation increases, it is possible that $\text{Im}\{Y_d\}$ or $\text{Im}\{Z_d\}$ might vary somewhat with V_0 or I_0 as well. When that occurs, the oscillation frequency changes as necessary so that the reactive part of (8.13) or (8.15) continues to be satisfied.

8.1.3.2 Stability Conditions

The description of the negative-resistance oscillator in the previous section assumes that the oscillation is stable. The idea of a stable oscillation seems at first to be an oxymoron, as the circuit must be unstable, in the linear sense, for oscillation to occur. Our concept of stability, in this case, is somewhat different; we call the oscillation *stable* if it returns to the steady-state conditions after being perturbed in some way.

Kurokawa derived expressions for the stability conditions as well. They are

$$\frac{\partial R_d}{\partial I_0}\frac{\partial X_L}{\partial \omega} - \frac{\partial X_d}{\partial I_0}\frac{\partial R_L}{\partial \omega} > 0 \tag{8.16}$$

$$\frac{\partial G_d}{\partial V_0}\frac{\partial B_L}{\partial \omega} - \frac{\partial B_d}{\partial V_0}\frac{\partial G_L}{\partial \omega} > 0 \tag{8.17}$$

where

$$\begin{aligned} Y_d &= G_d + jB_d \qquad & Z_d &= R_d + jX_d \\ Y_L &= G_L + jB_L \qquad & Z_L &= R_L + jX_L \end{aligned} \tag{8.18}$$

It is easy to see that (8.16) is satisfied in ordinary cases. For example, suppose that we have a series-resonant oscillator, consisting of a two-terminal negative resistance device and a lumped-element load, creating a series RLC resonance. The second term in (8.16) is clearly zero. Of the first term, $\partial R_d / \partial I_0 > 0$ and, for a series LC resonance, $\partial X_L / \partial \omega > 0$ as well.

8.1.4 Unified Treatment of Oscillation Conditions

It is possible to show that the Kurokawa (negative resistance) case and the Barkhausen case are equivalent, and to generalize both in the process. To do this, we generalize the Colpitts oscillator example from Section 8.1.2. Consider the multiport circuit shown in Figure 8.4. The device, with voltage **V**, generates a current **I**. We define these quantities in the pseudo-linear manner (8.12) to (8.14) on which Kurokawa theory is based. Then,

$$\mathbf{I} = \mathbf{Y}_d \mathbf{V} \tag{8.19}$$

where $\mathbf{Y}_d$ is the device admittance. The current excites the linear load network; for oscillation to occur, the resulting voltage must be **V**. This is a multiport analog of (8.7). Thus,

$$\mathbf{V} = -\mathbf{Z}_L \mathbf{I} \tag{8.20}$$

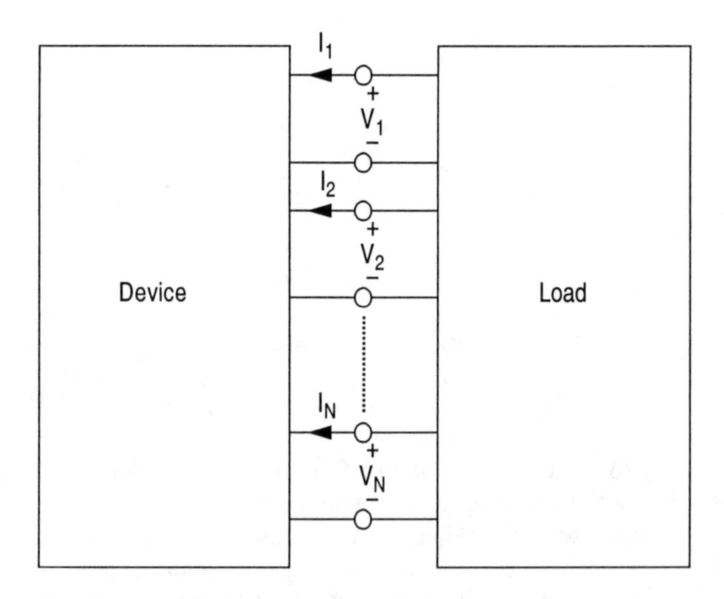

Figure 8.4 Generalized oscillator model, consisting of a device and load network. We assume that dc bias is part of the device. The load consists of the output port or ports and the device matching or embedding network.

where $\mathbf{Z}_L$ is the impedance matrix of the embedding circuit. Substituting for $\mathbf{I}$ gives,

$$-\mathbf{Z}_L\mathbf{Y}_d\mathbf{V} = \mathbf{V} \tag{8.21}$$

showing that the oscillation conditions are equivalent to saying that the matrix $-\mathbf{Z}_L\mathbf{Y}_d$ must have an eigenvalue equal to 1.0. Using scalars in (8.21) gives

$$Z_L Y_d + 1 = 0 \tag{8.22}$$

which yields the Kurokawa oscillation condition,

$$Y_L + Y_d = 0 \tag{8.23}$$

where $Y_L = 1 / Z_L$. We shall show, in Section 8.2.2, that a similar criterion satisfies an oscillator's harmonic-balance equation.

8.1.5 Linear Design

Long before nonlinear analysis of microwave oscillators was possible, design methods based on linear theory had been developed. These are still practical, by themselves, but also are part of many nonlinear analysis methods. We describe two such methods in this section. The first is to treat the oscillating device as a two-terminal negative resistance; the other is to define a feedback path and to satisfy (8.5). The former method is used most frequently for microwave circuits; the latter, for RF circuits.

8.1.5.1 Negative-Resistance Design

Even if we view a transistor as a two-port, it can be used in a negative-resistance design. To do this, we add feedback and terminate the output port with an appropriate load. This leaves the input port, which can be terminated with an impedance that satisfies the oscillation conditions. It is possible to show that, when oscillation conditions are satisfied at one port of a linear two-port, they are automatically satisfied at the second port.

Figure 8.5 shows two ways to create feedback for producing a negative resistance. In Figure 8.5(a), a capacitor has been connected to the emitter; in Figure 8.5(b), a base inductor is used. At some frequencies, or with certain devices, one of these configurations usually works better than the oth-

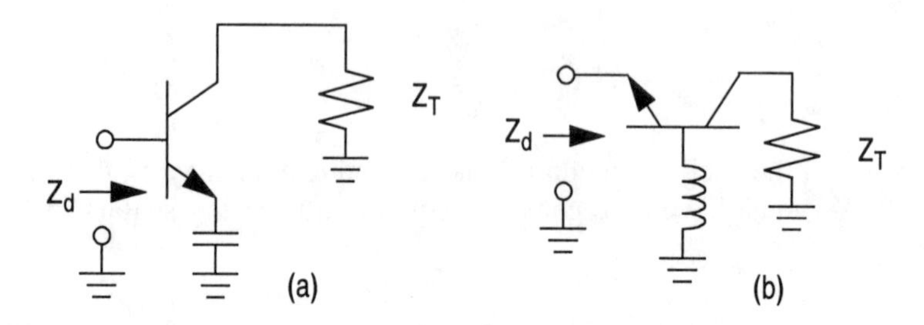

Figure 8.5 A transistor oscillator forms a two-terminal negative-resistance device when feedback is added and one port is terminated. (a) Capacitive feedback is used at the emitter. (b) Inductive feedback is employed at the base. The configuration used in any particular design is chosen for effectiveness in providing the required negative resistance and for practical considerations.

er; in certain cases, one may also be more practical. The goal is to create a negative real part of the device impedance, Z_d, or admittance, Y_d. This value should be approximately 100Ω for a series-resonant impedance or 0.01 S for a shunt resonance. The value is not critical, but unusually high or low impedances or admittances usually result in unreliable start-up.

Once we have achieved adequate negative resistance, the port must be terminated with a reactance or susceptance that resonates the device impedance or admittance. The termination need not have a resistive part, as the resistance/conductance of the port impedance/admittance decreases to zero (or, more precisely, a very small value that cancels the loss resistance of the termination) as the oscillation builds. This creates a dilemma, however, which can be illustrated by an example. Suppose, after optimizing the feedback capacitance or inductance, we determine that $Z_d = -100 + j50$. This means that we must use a capacitance having 50Ω reactance to series-resonate the port. However, suppose we viewed the port as an admittance; then $Y_d = -0.008 - j0.004$. This corresponds to a resonating capacitance having 250Ω reactance, and we parallel-resonate it. Which is correct?

The dilemma arises from the fact that we are not satisfying oscillation conditions exactly under small-signal conditions; if we were, our load would include a resistive part, and we could use either a series or shunt load network. The problem would then disappear. Instead, we expect the resistive/conductive part of the oscillator's port impedance to disappear as the oscillation grows, and we want to know what reactance is left after that happens. Therefore, we need to know whether the port is inherently a series or parallel resistance-reactance combination. This can be determined easily. If the real part of the port impedance is relatively constant with frequency, the port should be viewed as a series combination; if the real part of the admittance is constant, it should be viewed as a parallel one. The appropriate resonating element can then be selected easily.

In microwave oscillators, the resonating element can be realized in a number of ways; for example, it can be a lumped capacitance or inductance, a varactor, or a transmission-line stub. If the impedance of a desired tuning element (e.g., a varactor or dielectric resonator) does not satisfy oscillation conditions, it sometimes can be shifted by means of a transmission line. This approach is especially helpful in creating varactor-tuned oscillators, as the impedance of available varactors may not satisfy oscillation conditions, but a varactor terminating a transmission line, whose length is adjustable to provide an appropriate impedance, might work properly.

Dilemmas such as the one we have illustrated can be avoided through the use of nonlinear analysis of the oscillator, which provides not only a correct determination of the resonating impedance for a particular oscillator frequency, but also the output power and the phase-noise spectrum.

8.1.5.2 Feedback Design

One approach to oscillator design is to break the feedback loop, identifying the forward and feedback paths, and designing the oscillator to satisfy (8.6). Figure 8.6 shows how this can be accomplished. Figure 8.6(a) shows an oscillator consisting of an narrowband amplifier (which, in practice, consists of an amplifier and resonator), in which the input is connected to the output by a well-defined feedback path. The feedback path in Figure 8.6(a) corresponds to the path between the $F(s)$ block and the summing junction in Figure 8.1. The oscillator's load is not shown explicitly; it is part of the amplifier-resonator combination. Breaking the feedback loop gives us the equivalent circuit in Figure 8.6(b); that is, when $V_2 = V_s$, oscillation conditions are realized. This circuit can be set up easily in a linear circuit simulator, for example, in the AC analysis of SPICE. Of course, optimizing the circuit can be a little tricky, as any changes to the amplifier-resonator automatically change Z_{in}.

In microwave circuit simulators, however, all sources have source impedances, so we must use the circuit in Figure 8.6(c). Then, the voltage gain is

$$\frac{V_2}{V_1} = \frac{S_{21}}{S_{11} + 1}\sqrt{\frac{Z_L}{Z_s}} \tag{8.24}$$

where Z_s and Z_L are the source and load impedances, respectively, assumed to be real. An obvious choice is to make $Z_L = Z_s = Z_{in}$; then $V_2 / V_1 = S_{21}$.

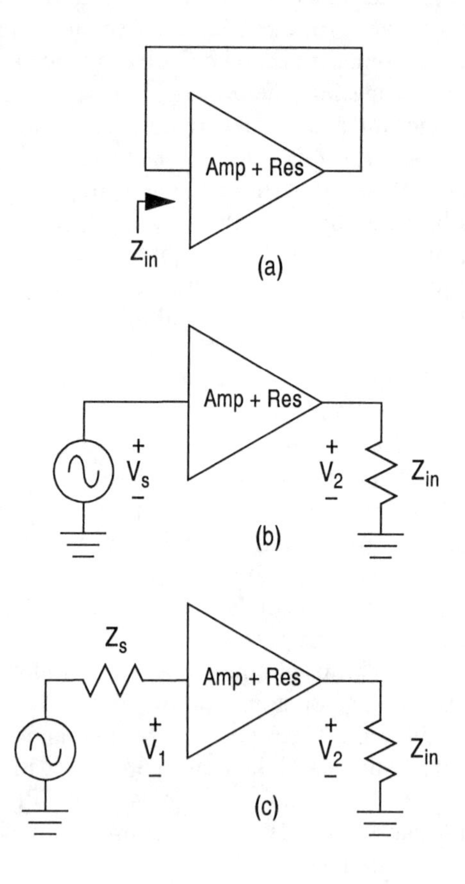

Figure 8.6 Breaking the feedback loop in an oscillator: (a) the oscillator, consisting of an amplifier/resonator and feedback; (b) equivalent circuit for oscillator design; (c) circuit configuration for use in a microwave circuit simulator.

An approach described by Rhea [8.2] realizes this method. The method is most useful for RF oscillators, satisfying (8.6) in a systematic and correct way. At microwave frequencies, the phase shift in the feedback path often makes the approach difficult to implement.

1. An amplifier is cascaded with a resonator as shown in Figure 8.7(a). Various types of resonators are possible, but the coupling to the resonator, and thus the loaded Q, should be adjustable in some way. The output port of the amplifier is not the output port of the oscillator; the oscillator load must be included as part of this circuit.
2. The circuit is designed according to the following criteria:

$$\begin{aligned} Z_s &= Z_L = Z_{in} \\ |S_{21}| &> 1.0 \\ \angle S_{21} &= 0 \end{aligned} \tag{8.25}$$

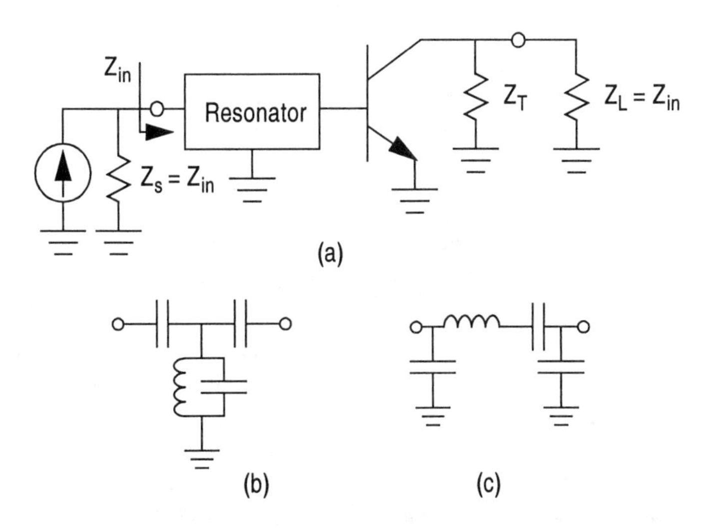

Figure 8.7 (a) Oscillator model for the feedback-oscillator design method described in Section 8.1.5.2. Z_T is the oscillator's output load. Once the conditions in (8.25) are satisfied, the output is connected to the input to create an oscillator. (b, c) Examples of resonators. The series capacitors in (b) and the shunt capacitors in (c) control the coupling of the resonator to the circuit, thus the loaded Q. The coupling should be as weak as possible consistent with satisfying (8.25).

where the impedances are real and S_{21} is evaluated at the frequency of maximum gain. These conditions can be most easily achieved by numerical optimization. The resonator Q should be made as high as possible, consistent with satisfying (8.25). In practice, $|S_{21}|$ should be at least 6 dB, preferably more.

3. The source and load are removed, and the output port is connected to the input port, to complete the oscillator design.

Although [8.2] does not propose it, this method could be extended into nonlinear design process. To do that, the linear amplifier should be replaced by its nonlinear model. Then, instead of the $|S_{21}|$ condition in (8.25), the drive level of the source is adjusted until the output voltage equals the input voltage. Other conditions in (8.25) remain the same. The frequency at which the conditions are satisfied and the power in Z_T, the oscillator load, are then the frequency and output power of the oscillator.

8.2 NONLINEAR ANALYSIS OF OSCILLATORS

The fundamental problems in nonlinear analysis of oscillators have been known for some time. These were covered in some detail in Section 6.4.1. In summary, the most serious are the following:

- The existence of a zero solution (i.e., a nonoscillating circuit);
- The lack of an inherent time reference;
- The fundamental frequency is unknown.

We also noted that the lack of a time reference implies that the harmonic-balance Jacobian is singular. That problem can be corrected by artificially inserting such a reference; however, even in that case, the Jacobian may still be seriously ill conditioned. In Section 6.4.1, these problems were discussed briefly, as necessary background for noise analysis. Here we examine practical methods for performing nonlinear oscillator analysis.

8.2.1 Simple Methods

8.2.1.1 Feedback Probe Method

This method is similar to the linear design method in Section 8.1.5.2, but the implementation is very different. In this case, the feedback loop is identified and a type of probe, similar to a directional coupler, is inserted into the loop. This coupler is used to inject a signal and to sample the fundamental-frequency feedback. The frequency and excitation level are adjusted until the large-signal oscillation conditions are met.

Figure 8.8 shows the circuit. The oscillator probe breaks the feedback circuit, allowing a generator to be inserted. The S parameters of the probe, at the fundamental frequency, are

$$S_{12} = S_{41} = S_{23} = 1.0 \tag{8.26}$$

At harmonic frequencies, they are

$$S_{12} = S_{21} = S_{41} = 1.0 \tag{8.27}$$

Otherwise, $S_{ij} = 0$. The port numbers of the probe are defined in the figure.

The probe element effectively breaks the circuit at the fundamental frequency, in the forward direction. The source then replaces the forward-propagating wave in the loop; adjusting it so that $V_2 = V_s$ (in both magnitude and phase) satisfies the oscillation conditions.

In these methods, the tricky part of the process is to identify the feedback loop. This requires some judgment on the part of the user. In many types of circuits, the feedback path may not be obvious; then, an error is

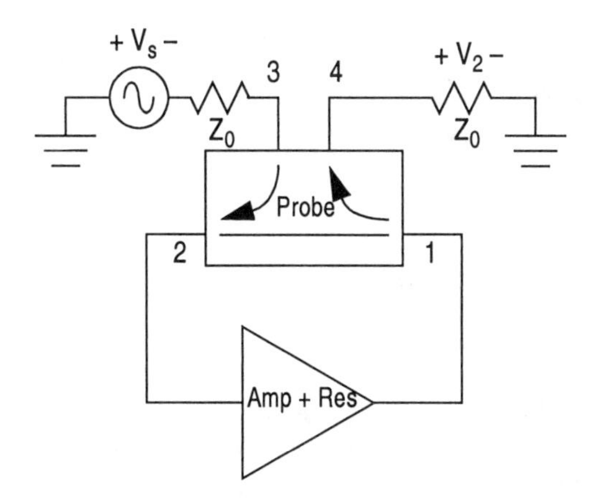

Figure 8.8 Feedback probe method. The probe breaks the feedback loop in the forward direction only, at the fundamental frequency.

likely. Harmonic balance methods that solve the oscillator equations without resorting to such artificial means therefore have a significant advantage.

8.2.1.2 Device-Line Method

Device-line measurement is a clever idea introduced by Wagner in 1979 [8.3]. It was developed to address an obvious problem with the linear oscillator-design methods in use at the time: those methods cannot optimize the output power. Wagner proposed a measurement on the output port of the oscillator that would solve this problem. The method, although originally intended as a measurement technique, can easily be implemented in a nonlinear simulator. As such, it becomes a design technique.

To optimize the output power, we must determine the load impedance. Ideally we would like to attach a tuner to the output of the oscillator, vary the load impedance, and plot power versus impedance contours on a Smith chart. With computer-controlled load-pull tuners available today, this is possible; however, in 1979 such tuners did not exist. Other methods were needed.

The underlying reasoning is as follows: suppose we terminate the oscillator in an impedance that does not allow oscillation, and we apply power to the output port. Because the port has a negative impedance, the oscillator applies power to the source; that power is dissipated in the source termination. We hypothesize that there is no difference between free running oscillation that produces a certain voltage and current at the output port, and the same voltage and current created by the generator. At least in terms of the fundamental frequency, this reasoning is valid. Then, if V and I are the fundamental-frequency voltage and current at the port, we can calculate

$$\begin{aligned} Z_L &= \frac{V}{I} \\ P_o &= -\frac{1}{2}\mathrm{Re}\{VI^*\} \end{aligned} \tag{8.28}$$

where P_o is the output power.

This set of conditions is easy to create in any harmonic-balance simulator. Because of the negative resistance at the port, however, convergence can be tricky. Using a very high or very low source impedance can be helpful to provide good convergence. For a series-resonant port, a high-impedance source is needed; for a parallel resonance, a low impedance.

The generator may require fairly high power; this creates no problem in the analysis, as the generator power has no physical significance.

A problem in this approach is that all harmonics are terminated in the load impedance used in the measurement, not the synthetic impedance defined by (8.28). This load is usually an unrealistically high or low impedance. The auxiliary generator method, described in Section 8.2.2.4, is similar in concept, but avoids this difficulty.

8.2.2 Harmonic-Balance Analysis of Oscillators

Because the noise analysis of oscillators often depends on the results of their harmonic-balance analysis, we must first consider the large-signal analysis of those circuits. We must modify the analysis from the classical one introduced in Section 6.1.1, addressing the three basic problems described above: the zero solution, lack of a time reference, and determination of the oscillation frequency. Many methods have been developed to address these needs; we describe a few of them in this section.

8.2.2.1 Basic Formulation

As with all nonlinear circuits, the oscillator must satisfy the harmonic-balance equation, (6.7), illustrated by Figure 6.1. This is merely a statement of Kirchoff's current law, evaluated at the interface between the linear and nonlinear subcircuits. That expression, repeated here for convenience, is

$$\mathbf{F}(\mathbf{V}) = \hat{\mathbf{Y}}_{1,1}\mathbf{V} + \mathbf{I}_s + \mathbf{I}_{\mathrm{NL}} = \mathbf{0} \tag{8.29}$$

where $\mathbf{I}_{\mathrm{NL}}$ is the current in the nonlinear subcircuit, $\mathbf{V}$ is the set of interface voltage components, and $\hat{\mathbf{Y}}_{1,1}$ is a submatrix of $\mathbf{Y}$, the admittance matrix of the linear subcircuit. In nonautonomous circuits, $\mathbf{I}_s$ is the vector of excitations, transformed to current sources at the ports connecting the linear and nonlinear subcircuits. In such circuits it includes RF and dc excitations. In an autonomous circuit, however, there are no RF excitations so $\mathbf{I}_s$ includes only dc excitations. To simplify things even more, we can imagine the dc excitations to be inherently part of the nonlinear elements; that is, they are biased, active devices. This assumption exists implicitly in Kurokawa theory, and it is reasonable to employ it in the general case. Then, (8.29) becomes

$$\mathbf{F}(\mathbf{V}) = \hat{\mathbf{Y}}_{1,1}\mathbf{V} + \mathbf{I}_{\mathrm{NL}} = \mathbf{0} \tag{8.30}$$

Note that this is entirely consistent with the generalized Kurokawa case described in Section 8.1.4. Multiplying both sides of (8.30) by $\hat{\mathbf{Z}}_{1,1} = \hat{\mathbf{Y}}_{1,1}^{-1}$ gives

$$-\hat{\mathbf{Z}}_{1,1}\mathbf{I}_{NL}(\mathbf{V}) = \mathbf{V} \tag{8.31}$$

which is entirely analogous to the expression derived for the Kurokawa case,

$$-\mathbf{Z}_L\mathbf{Y}_d\mathbf{V} = \mathbf{V} \tag{8.32}$$

where $\mathbf{Y}_d\mathbf{V}$ was the current in the nonlinear device, equivalent to $\mathbf{I}_{NL}$ in the harmonic-balance case.

Although many of the design methods described earlier focussed only on the oscillator's operation at the fundamental frequency, this expression implies that the oscillation conditions must be satisfied at all harmonics, not just the fundamental. This is a simple implication of the need to satisfy Kirchoff's laws throughout the circuit. We shall see, furthermore, that the use of an inadequate number of harmonics (specifically, just the fundamental) can give erroneous results in phase-noise calculations.

8.2.2.2 Frequency-Component Replacement

It seems essential that, one way or another, we create a time reference for the circuit waveforms. A brute-force method for doing this is simply to set the imaginary part of one significant fundamental-frequency voltage component to zero. That component can then be eliminated, along with the column of (6.1) that multiplies it. The result is an overspecified system of equations in (6.7); that is, there is one more equation than unknown. Such systems can be solved in a number of ways; the simplest, in concept, is to determine an equation (corresponding to a matrix row) that is most nearly linearly dependent on the others, and simply to eliminate it. There is no great penalty in a little sloppiness in making the decision, since Newton's method is based on an approximation of the zero at each iteration, and thus is relatively tolerant of small errors.

A clever approach is to replace the deleted component with another quantity. (We examined this method briefly in Section 6.4.4.) For oscillator analysis, one can use the frequency of oscillation. Then, the oscillation frequency is found naturally as part of the harmonic-balance analysis. The sensitivity of each current component to the frequency, which is directly useful in phase-noise analysis, is also determined when the Jacobian is formed.

Conversely, it is possible to fix the oscillation frequency and replace the deleted component with a circuit parameter. The circuit parameter is then varied in the solution process to guarantee oscillation. The circuit parameter is usually one that can be trusted to affect the oscillation frequency strongly; for example, the length of a resonant stub or the capacitance of a tuning varactor. In this way, the analysis process actually becomes a type of synthesis procedure, in which the oscillator is designed to operate at a specific frequency.

In either case, the Jacobian must be modified from the usual harmonic-balance formulation. The replacement quantity, whether a frequency or circuit parameter, is likely to be very different in magnitude from the other Jacobian entries. Therefore, to preserve good conditioning of the matrix, it is wise to scale that quantity appropriately, so the magnitude of the entry is on the same order as other Jacobian entries.

8.2.2.3 Optimization

The oscillator analysis can be placed in an optimization loop. Although this adds a layer of complexity to the analysis and increases the computational cost, it can be very effective.

If the optimization goals are carefully chosen, optimization can help to avoid the undesired zero solution. One possible method is to invoke the Kurokawa condition as an optimization goal:

$$Y_{\mathrm{NL}}(\omega_p) + Y_L(\omega_p) = 0 \tag{8.33}$$

where $Y_{\mathrm{NL}}(\omega_p)$ is the fundamental-frequency large-signal admittance at one significant nonlinear element port in the analysis, defined as

$$Y_{\mathrm{NL}}(\omega_p) = \frac{I(\omega_p)}{V(\omega_p)} \tag{8.34}$$

and $Y_L(\omega_p)$ is the linear embedding admittance at that port. From (8.34) we can see that, as long as $Y_L(\omega_p)$ is nonzero, (8.33) cannot be satisfied by the zero solution. The choice of a variable of optimization is somewhat arbitrary, since satisfying (8.33) in the harmonic-balance process guarantees that (8.30) is satisfied completely. It seems that $V(\omega_p)$ would be a logical choice.

8.2.2.4 Auxiliary Generator

The use of a fictitious signal generator, which we call an *auxiliary generator*, can be very helpful in eliminating the problems of the zero solution and undefined time reference [8.4, 8.5]. It also improves the conditioning of the Jacobian matrix, improving the reliability of convergence in harmonic-balance analysis.

The idea behind the auxiliary generator is illustrated in Figure 8.9. The generator is a type of probe, consisting of a sinusoidal voltage source and series impedance or current source and shunt admittance. The impedance is zero (or some small value) at the generator's frequency and infinite at all other frequencies. The magnitude and frequency of the source is adjusted by an optimization loop; its phase is always zero. Similarly, the current-source probe has zero shunt admittance at the generator frequency and infinite admittance at other frequencies. Like the voltage source, its frequency and magnitude are adjusted by an optimization loop.

The source chosen for any particular oscillator analysis depends on the oscillator's circuit structure at the point at which the generator is connected. The type of probe and location of the probe connection is difficult to generalize; these require a degree of judgement. The oscillator's resonator is usually a good place to connect the probe. If a parallel resonator is used, the voltage probe should be connected in parallel with it; for a series resonance, the current probe, connected in series, should be employed. Other good locations for a voltage probe are the base-emitter junction of a BJT or the gate-source of a FET.

The auxiliary generator is connected to an appropriate point in the oscillator circuit and a nonlinear analysis is performed. The probe's voltage or current and frequency are adjusted, and the harmonic-balance analysis is repeated, until no current is detected in the series impedance, or no voltage appears across the shunt admittance. Then, the source can be removed without changing the operation of the circuit, and the voltage/current waveforms in the circuit are those of the free-running oscillator. (The voltage-source auxiliary generator is simply removed; the current-source auxiliary generator is replaced by a short circuit.)

The operation of the optimization loop is critical to the success of the analysis. Simply treating the source voltage/current and frequency as variables of optimization is usually not sufficient to guarantee success. In oscillators having high-Q resonators, optimization alone is often too coarse to find a precise solution. It is usually necessary to implement some kind of

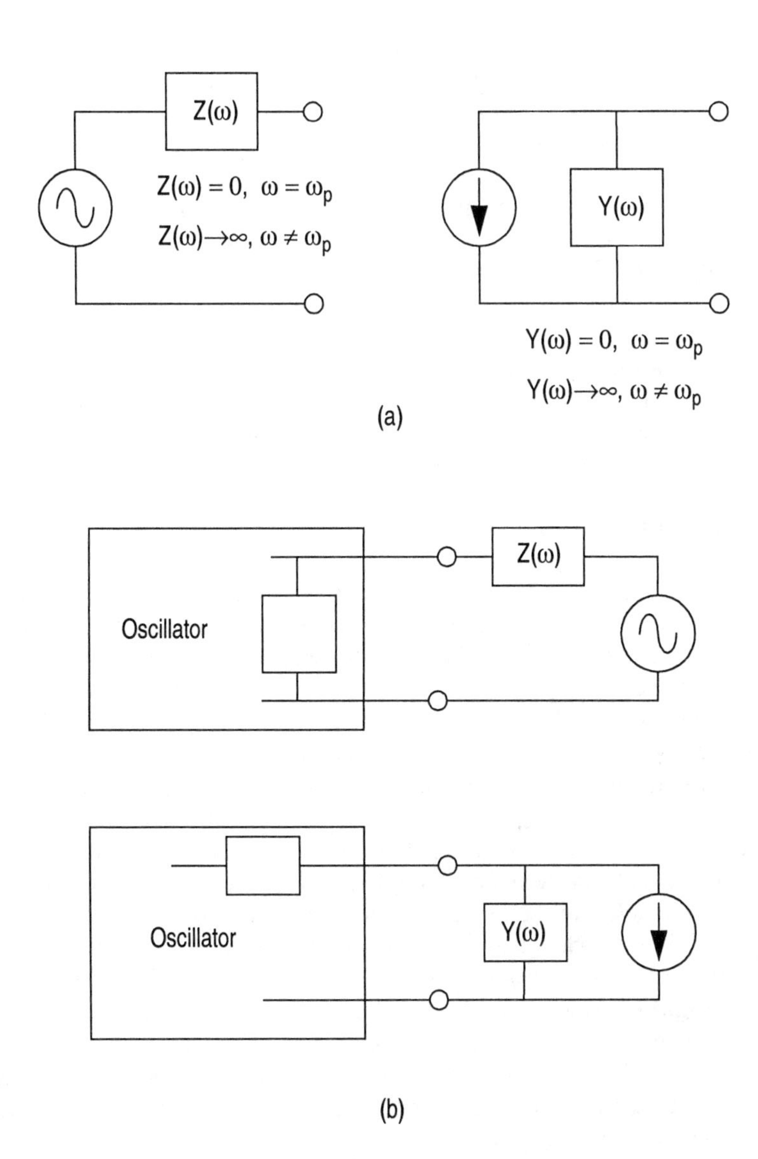

Figure 8.9 (a) Auxiliary generators used for oscillator analysis. (b) The generators are connected to a shunt or series element in the oscillator, typically a resonator.

direct search to locate a range in which a solution is likely; as a minimum, it must find a region that satisfies the linear oscillation conditions, described in Sections 8.1.3 and 8.1.5, under small-signal conditions. The optimizer then can complete the task, gradually increasing the source voltage while following the resulting change in the resonant frequency.

This method clearly prevents a zero solution and introduces a time reference. The optimization loop controls the frequency, so all three of the major analytical difficulties are eliminated. An important characteristic of this method is its use of a generator, which substantially improves the numerical conditioning of the Jacobian, leading to strong convergence to a solution in the harmonic-balance analyses.

The use of an auxiliary generator is not unprecedented. It can be viewed as a generalization of the device line measurement technique, translated into numerical form (Section 8.2.1.2 and [8.4, 8.6]).

8.2.3 Time-Domain Analysis

Oscillators can be analyzed effectively by time-domain analysis. Time-domain analysis can provide important information about oscillators that is difficult to obtain by harmonic balance; for example, the start-up transient of a fixed-frequency oscillator or the transient response of a voltage-controlled oscillator (VCO) to a step change of frequency-control voltage.

Time-domain analysis inherently solves the problems of the time reference and frequency, as these are created naturally as part of the analysis. It is subject to the zero-solution problem, however, and ill conditioning of the relevant matrices may also be troublesome. To obtain an oscillatory response, it is usually necessary to provide some kind of start-up transient. Common approaches are to begin the analysis with a voltage spike at some appropriate circuit node, or to turn on the dc power with a step function.

Oscillators using multiple transistors, such as frequency doubling oscillators or multivibrators, may require the artificial introduction of a minor imbalance or asymmetry to start the oscillation. This can be as simple as making the transistors' current gains or transconductances a few percent different. The start-up transient, in this case, may be sensitive to the imbalance, so the process should be approached with care.

Oscillators, like many RF and microwave circuits, often contain a mix of long and short time constants. Long time constants are generally associated with the bias circuitry. When long time constants exist, it may be necessary to integrate the oscillator equations for an uncomfortably long time before steady-state conditions are reached. Some time-domain software includes methods, such as shooting methods, that help to determine steady-state conditions without the need for such long integrations.

8.3 NOISE IN OSCILLATORS

It has long been recognized that noise in an oscillator circuit causes phase fluctuations. These random phase deviations are manifest as a noise spectrum mirrored on each side of the oscillator frequency. Such phase noise is troublesome in any system in which a signal's phase carries information; for example, in analog frequency- or phase-modulation, or digital phase- or phase-amplitude modulation. Because the mixers in such systems are essentially phase adders, they transfer the oscillator's phase noise, degree for degree, to the received signals.

An oscillator's phase-noise spectrum is not white. It usually has a slope, close to the carrier, of 30 dB/decade of offset frequency. 20 dB/decade of the slope comes from the inherent resonance in the oscillator, which sets its frequency, and an additional 10 dB/decade comes from $1/f$ noise, upconverted from baseband by the oscillating device's nonlinearities, which is the dominant contributor.

8.3.1 Leeson's Model

An early model of oscillator noise by Leeson [8.7] has been cited frequently to describe the effect of oscillator parameters, especially resonator Q, on phase noise. Leeson used a phase-feedback model of an oscillator, consisting of an amplifier, resonator, and noise source, to develop an expression for phase noise. The feedback system formulated at the oscillator frequency is converted to a baseband equivalent circuit.

The model is based on the circuits in Figure 8.10. Figure 8.10(a) shows the oscillator modeled as a feedback structure consisting of an amplifier and resonator. In this sense, it is much like the treatment in Section 8.1.5.2. Noise is treated as an additive source at the input of the amplifier. We assume that the amplifier's gain is flat within the narrow band of interest near the oscillation frequency.

Since there is nothing special about the oscillation frequency, this model can be shifted to baseband and described in terms of the modulation—in this case, the noise—alone. The equivalent circuit is shown in Figure 8.10(b). At baseband, the resonator becomes a low-pass circuit. Its transfer function, $T(f_m)$, is

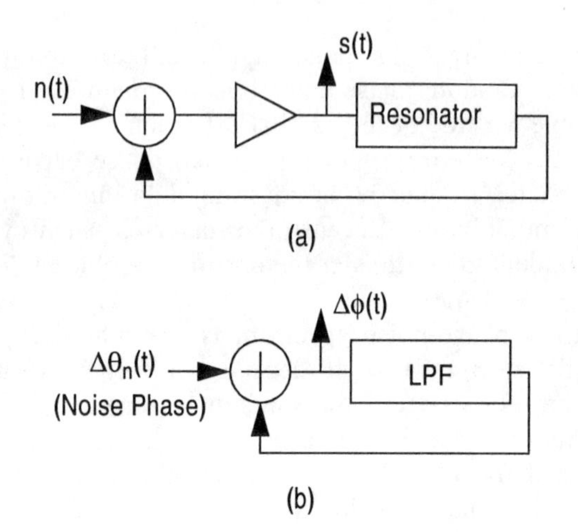

Figure 8.10 Leeson's model of a noisy oscillator: (a) oscillator model; (b) equivalent phase-feedback loop.

$$T(f_m) = \frac{1}{1 + j2Q_L\dfrac{f_m}{f_0}} \tag{8.35}$$

where f_m is the baseband frequency (equal to the offset frequency of the noise), f_0 is the output frequency, and Q_L is the loaded Q of the resonator. $\Delta\theta_n(t)$, in the figure, is the phase component of the noise; $\Delta\phi(t)$ is the resulting phase deviation of the oscillator signal. The RMS phase deviation of $\Delta\theta_n(t)$ is

$$\Delta\theta_{n,\,\mathrm{RMS}} = \operatorname{atan}\left(\frac{V_{n,\,\mathrm{RMS}}}{V_s}\right) \approx \frac{V_{n,\,\mathrm{RMS}}}{V_s} \tag{8.36}$$

where $V_{n,\,\mathrm{RMS}}$ is the double-sideband RMS noise voltage and V_s is the signal voltage. The approximation is very good, since $V_{n,\,\mathrm{RMS}} \ll V_s$. This is called the *small-phase approximation,* which is employed universally in phase-noise analysis.

Leeson expresses the mean-square phase noise as

$$\overline{|\Delta\theta_n|^2} = \frac{FKT_0}{P_s} \tag{8.37}$$

where F is a noise figure, K is Boltzmann's constant, and T_0 is the standard temperature of 290K. The noise includes the ever-present 290K termination noise. This expression has engendered 40 years of confusion, as it implies that $n(t)$ is a white noise source and equal to the high-frequency noise of the amplifier. In fact, it is upconverted $1/f$ noise, which is distinctly non-white and much greater in magnitude than the amplifier's high-frequency noise. A better expression might be

$$\overline{|\Delta\theta_n|^2(f_m)} = \frac{S_n(f_m)}{P_s} \tag{8.38}$$

where $S_n(f_m)$ is the noise power spectrum.

From ordinary feedback theory, the transfer function is

$$\Delta\phi = \frac{1}{1 - T(f_m)}\Delta\theta \tag{8.39}$$

and substituting (8.35) and squaring gives

$$S_\phi(f_m) = \overline{|\Delta\phi|^2(f_m)} = \left[1 + \frac{f_0^2}{f_m^2}\frac{1}{4Q_L^2}\right]\overline{|\Delta\theta_n|^2(f_m)} \tag{8.40}$$

The upconverted phase noise has the spectrum,

$$S_n(f_m) = P_n\left(1 + \frac{f_c}{f_m}\right) \tag{8.41}$$

where f_c is called the *corner frequency* of the noise and P_n is a constant. In terms of the original, unfortunate notation, $P_n = FKT_0$. Finally, substituting in (8.40) gives

$$S_{\phi}(f_m) = \frac{P_n}{P_s}\left(1 + \frac{f_c}{f_m}\right)\left[1 + \frac{f_0^2}{f_m^2}\frac{1}{4Q_L^2}\right] \tag{8.42}$$

Usually we wish to know the noise-to-carrier power ratio, which is called $L(f_m)$. Simple phase-modulation theory [8.8] gives

$$L(f_m) = \frac{1}{2}S_{\phi}(f_m) = \frac{P_n}{2P_s}\left(1 + \frac{f_c}{f_m}\right)\left[1 + \frac{f_0^2}{f_m^2}\frac{1}{4Q_L^2}\right] \tag{8.43}$$

The above expression, known as *Leeson's equation*, predicts an $L(f_m)$ spectrum that varies 20 dB/decade within the bandwidth of the resonator, caused by the oscillator, and an additional slope of 10 dB/decade at offset frequencies below the noise corner f_c, caused by the spectrum of the noise itself. This creates several regions in $L(f_m)$ with distinct slopes, shown in Figure 8.11. The slopes depend on the relationship between the noise corner frequency f_c and the Q of the oscillator's resonator. In all cases, the slope close to the frequency of oscillation is 30 dB/decade. If the Q is low, so $f_0/2Q_L \gg f_c$, the region where $f_c < f_m < f_0/2Q_L$ has a 20-dB slope, caused by the combination of the white-noise part of $S_n(f_m)$ and the oscillator's resonance. If the Q is high, so $f_0/2Q_L \ll f_c$, the region $f_0/2Q_L < f_m < f_c$ has a 10-dB/decade slope, caused by $1/f$ noise outside

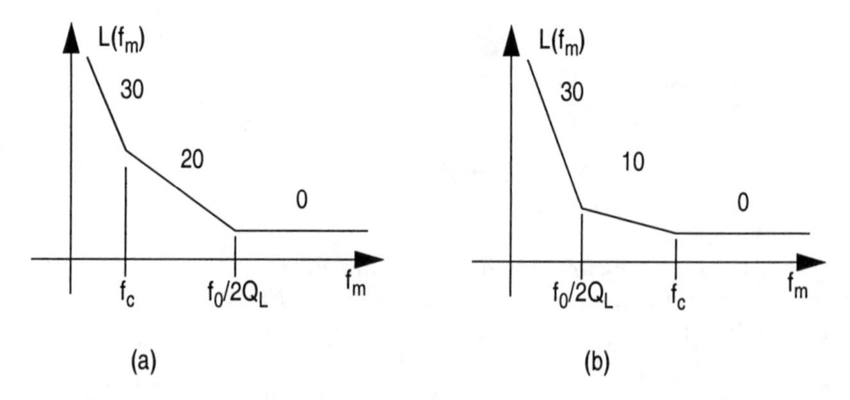

Figure 8.11 Oscillator carrier-to-noise spectra predicted by Leeson's model: (a) low-Q oscillator; (b) high-Q oscillator.

the resonator's bandwidth. At large offsets, $S_n(f_m)$ is flat and the oscillator's resonance has no effect on the noise. Then, phase noise consists of the phase component in additive noise plus a sinusoid, so the spectrum is flat.

In spite of its widespread use and fundamental validity, Leeson's model has a couple of problems that we should note. First, it tells us nothing about the noise upconversion process, which is critical to a quantitative treatment of oscillator noise. It merely assumes that the noise exists and can be treated as an additive source at the input of the amplifier. This is, fundamentally, a system model, not unlike the noise models we treat in Chapter 3. As such, it says little about the noise process in the circuit itself. Second, the concept of the loaded Q, Q_L, is difficult to translate to other kinds of oscillators. In Leeson's model, where we deal with a bandpass structure, the loaded Q is easy to define: it is simply $f_0/\Delta f$, where Δf is the 3-dB bandwidth. In the negative-resistance model, however, the real parts of the shunt admittances or series impedances cancel, creating a resonator having infinite Q. In that case, we find that the ratio C/L, in the parallel-resonant case, or L/C, in the series-resonant case, should be viewed in the same way as Q_L in Leeson's model. In microwave resonators, which use distributed elements, the Q can be defined in terms of slope parameters,

$$\chi = \frac{\omega_0}{2}\frac{dX}{d\omega}\bigg|_{\omega = \omega_0} \tag{8.44}$$

for series resonances, and

$$\xi = \frac{\omega_0}{2}\frac{dB}{d\omega}\bigg|_{\omega = \omega_0} \tag{8.45}$$

for parallel resonances. These can be equated, close to ω_0, to LC resonators. For series and parallel resonances, the inductance and capacitance of equivalent LC resonators are, respectively,

$$\begin{aligned} L &= \frac{\chi}{\omega_0} \\ C &= \frac{\xi}{\omega_0} \end{aligned} \tag{8.46}$$

Even with these clarifications, Leeson's model has limited quantitative value. Its value lies primarily in illustrating qualitatively how the inherent noise of the device, combined with the Q of the oscillator's resonance, affects the phase-noise spectrum of the oscillator.

8.3.2 Devices for Oscillators

The need to generate stable sinusoidal signals has been paramount throughout the history of electronic communications. Early radio oscillators used vacuum tubes, often in configurations (such as Hartley and Colpitts oscillators) that are still used, with transistors, today. Microwave tubes, such as klystrons, magnetrons, and backward-wave oscillators, eventually gave way to smaller, simpler, and more reliable solid state devices for most low-power applications. The first of these were two-terminal diode oscillators using Gunn, IMPATT, or tunnel-diode devices. Gunn devices and tunnel diodes had inherent negative resistance, while IMPATTs created negative resistance through transit time effects. All these devices had significant disadvantages: tunnel diodes could not generate even modest power levels; Gunns and (especially) IMPATTs were noisy, difficult to tune, and, in some cases, unreliable. Some of these devices remain in occasional use, usually at millimeter wavelengths.

Today, most oscillators are realized with transistors. At lower frequencies (below a few gigahertz), silicon homojunction devices are most often used. At higher frequencies, HBTs and microwave FETs are more common. Still, advanced silicon devices, including SiGe HBTs and CMOS devices, with cutoff frequencies in the millimeter range, are maturing and present real possibilities for oscillator use in the microwave and perhaps even millimeter-wave regions.

8.3.2.1 Bipolar Devices

Generally, bipolar devices are preferred for oscillators, as they have lower levels of $1/f$ noise than FETs. This results in lower phase noise. Commonly available silicon homojunction BJTs are often used at frequencies up to approximately 10 GHz, with appropriate packaging, and some advanced devices have been used up to 40 GHz. Silicon-germanium HBTs retain the low cost of mature silicon IC technologies, while providing operation into the upper microwave frequency range. SiGe HBT technology is advancing rapidly at this writing; the limitations of this technology are not yet visible.

For moderate cost oscillators at high frequencies, HBT devices in III-V technologies are preferred. InP HBTs have demonstrated cutoff frequencies of 200 GHz and more, while lower-cost, prosaic InGaP technologies, with cutoff frequencies of 50 to 100 GHz, have become the workhorses of the cellular and wireless LAN industries.

Most HBT processes are used for integrated circuits. Only a few manufacturers offer discrete HBTs.

8.3.2.2 FET Devices

FETs are usually used at frequencies where adequate bipolars are not available, or when system constraints dictate the use of a FET IC technology. Although microwave FETs have lower levels of high-frequency noise than homojunction bipolars or HBTs, their levels of $1/f$ noise are higher. As a result, they exhibit phase-noise levels on the order of 10 dB greater than bipolars. The high level of $1/f$ noise is generally attributed to imperfections in the crystal near the boundaries of the channel.

On the other hand, the high cutoff frequencies of FET devices allow them to be used as oscillators at frequencies considerably higher than bipolars. Modern HEMT devices can be used as oscillators well into the millimeter range and have largely supplanted both GaAs and InP Gunn devices in such applications.

8.3.3 Noise Sources and Modeling

Noise models of solid-state devices are described in Chapter 4. It is interesting to note that the SPICE noise models are still the dominant ones used for nonlinear noise analysis. There are a few exceptions; for example, the Chalmers models and BSIM include the results of more recent modeling work.

For both FETs and bipolars, the $1/f$ noise sources dominate in establishing oscillator phase noise levels. In FETs, a single source, in parallel with the channel, invariably is used. Bipolars have two sources, one in parallel with the base-to-emitter (BE) junction and another in parallel with the base-to-collector. In normal operation, only the BE junction conducts. In large-signal circuits, such as oscillators and power amplifiers, it is possible for both junctions to conduct, at some point in the waveform, and generate significant noise.

The general expression for $1/f$ noise given by (2.54), is used for both FETs and bipolars. Some implementations in versions of SPICE and other circuit simulators modify the basic expression somewhat; for example, to allow regions of the spectrum with different slopes or, as in the Chalmers model, more complex dependence on drain current. BSIM3 allows the use of the classical SPICE model, but also offers an extended model using two separate terms (4.32).

Because $1/f$ noise is low frequency, measuring the noise level of the devices and fitting it to the appropriate expression is relatively easy. The noise level is measured by a low-frequency spectrum analyzer at a number of bias values, the effect of resistive parasitics is removed, and the resulting spectra are fit to the appropriate expressions.

8.3.4 Heuristic Example: Van der Pol Oscillator

The Van der Pol oscillator is a simple oscillator circuit, which, unlike most oscillators, can be treated exactly by analytical means. It is therefore useful to impart an intuitive sense of the operation of an oscillator. It can also be used to provide insight into oscillator noise characteristics.

The Van der Pol is fundamentally a negative-resistance oscillator. It consists of a parallel LC resonator, load conductance, and nonlinear element. The equivalent circuit is shown in Figure 8.12. The nonlinear element's I/V characteristic is

$$I(V) = -GV + G\alpha V^3 \tag{8.47}$$

where G and α are parameters, $G > 0$. Figure 8.13 shows an example of the I/V characteristic when $G = 0.015$ and $\alpha = 1/3$; the negative-resistance characteristic is obvious. The inductor and capacitor are resonant at the oscillator frequency, so only a fundamental-frequency component of voltage exists across the nonlinear element. The current in the nonlinear element is, of course, nonsinusoidal, but the harmonics beyond the first simply idle in the resonator. The voltage across the load, $V(t)$, is

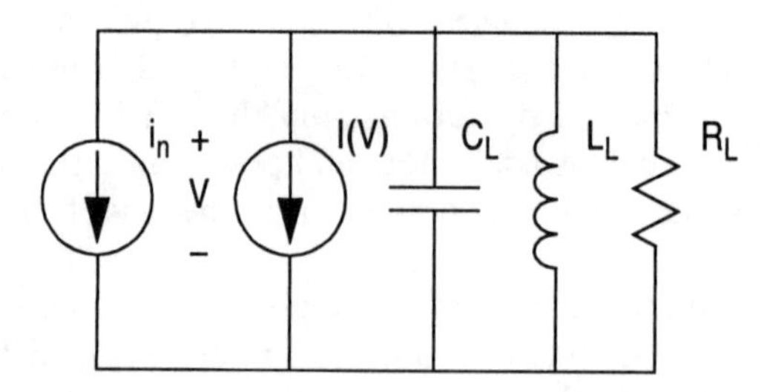

Figure 8.12 Equivalent circuit of the Van der Pol oscillator. The current source $I(V)$, given by (8.47), is nonlinear. The noise source, i_n, is assumed to be baseband $1/f$ noise.

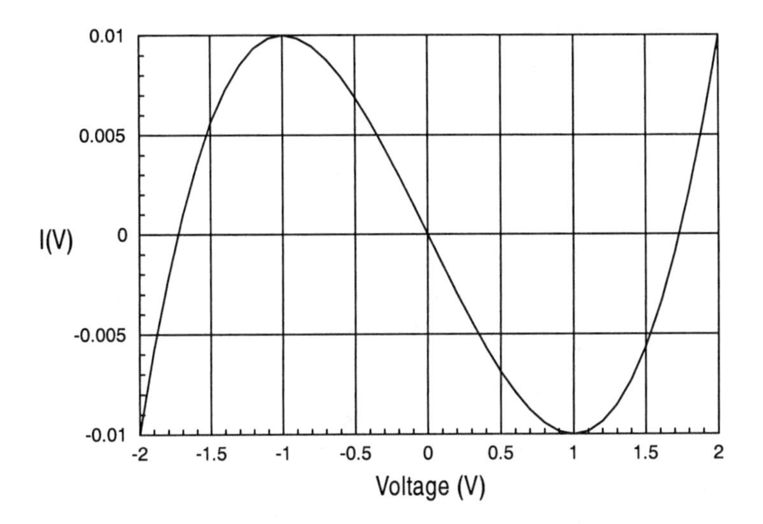

Figure 8.13 *I / V* characteristic of the nonlinear device in the Van der Pol oscillator example. $G = 0.015$, $\alpha = 1/3$.

$$V(t) = V_1 \cos(\omega t) \tag{8.48}$$

The large-signal oscillation condition becomes

$$G_L V_1 = -I_1 \tag{8.49}$$

where I_1 is the fundamental-frequency component of the current in the nonlinear element. Substituting (8.48) into (8.47) gives

$$I(t) = \left(-GV_1 + \frac{3}{4}G\alpha V_1^3\right)\cos(\omega t) + \frac{1}{4}G\alpha V_1^3 \cos(3\omega t) \tag{8.50}$$

To satisfy (8.49) we need

$$G_L V_1 = -\left(-GV_1 + \frac{3}{4}G\alpha V_1^3\right) \tag{8.51}$$

We concern ourselves only with the fundamental-frequency component of the voltage, as higher harmonics are short-circuited by the resonator, so only a fundamental-frequency voltage and current exist in the load. Equation (8.51) is solved to obtain

$$V_1 = \sqrt{\frac{4}{3}\left(\frac{G - G_L}{G\alpha}\right)} \tag{8.52}$$

Figure 8.14 shows the current waveform in the oscillating device of Figure 8.13 when $G_L = 0.01$. In order to have negative conductance at start-up, we must have $G_L < G$; then, the device admittance, defined as in (8.12), decreases as the oscillation builds, until it finally satisfies the oscillation conditions at the steady state. It is interesting to see how this happens. As the oscillation builds (i.e., as V_1 increases), the cubic term in (8.50) slows the increase in the fundamental-frequency current, causing the device admittance (defined in the Kurokawa sense) to decrease. From (8.52), when the peak voltage reaches $\sqrt{4/3} = 1.1547$ V, the device admittance equals G_L precisely. In this case, the peak voltage is somewhat greater than the voltage at the peak-current point of the I/V characteristic, 1.0V. If the voltage were to increase further, the fundamental-frequency current component

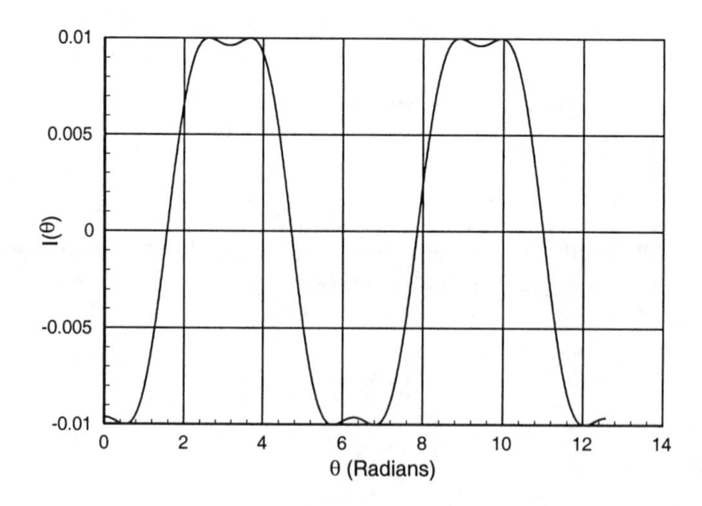

Figure 8.14 Current waveform in the nonlinear device in the oscillator example. $G_L = 0.01$. Other parameters are those in Figure 8.13.

would actually decrease (since we have coincidentally chosen the load that provides maximum fundamental-frequency current), and the device admittance would decrease further.

To illustrate the problems in using conversion-matrix analysis in autonomous circuits, we apply it to this problem in the hope of determining the noise level. First, to create a conversion matrix, we need the derivative waveform. The derivative of (8.47), $G_d(V)$, is

$$(G_d)(V) = -G(1-3\alpha V^2) \tag{8.53}$$

Substituting (8.48) and (8.52) into (8.53) and adding G_L gives the total conductance, $G(t)$:

$$G(t) = G_d(t) + G_L = G - G_L + 2(G - G_L)\cos(2\omega t) \tag{8.54}$$

from which we obtain the harmonic components,

$$G_0 = G - G_L \qquad G_1 = 0 \qquad G_2 = G_{-2} = G - G_L \tag{8.55}$$

All higher harmonics are zero. The *complete* conversion matrix becomes

$$\mathbf{G}_c = \begin{bmatrix} G_0 + Y^*(\omega_{-1}) & 0 & G_{-2} \\ 0 & G_0 + Y(\omega_0) & 0 \\ G_2 & 0 & G_0 + Y(\omega_1) \end{bmatrix} \tag{8.56}$$

where $G_0 = G_2 = G_{-2} = G - G_L$. $Y(\omega_j)$ is the admittance of the resonator at the frequency ω_j, an offset frequency. We use the notation in Chapter 6; thus, ω_0 is baseband, ω_1 is the upper sideband, and ω_{-1} is the lower sideband.

In (8.56) we can see that the matrix is severely ill conditioned and obviously singular at zero offset. The quantity $Y(\omega_0) \to \infty$, the admittance of the resonator at the baseband frequency, is enough to guarantee this. However, even in the absence of the resonator, the matrix representing the nonlinear device alone is still singular. The conversion matrix is ill conditioned, even though (8.56) is derived with a specific time reference.

The noise voltage correlation matrix, $\mathbf{C}_v$, is

$$\mathbf{C}_v = \mathbf{Z}_c \mathbf{C}_i \mathbf{Z}_c^{*\mathrm{T}} \tag{8.57}$$

Inverting (8.56) gives

$$\mathbf{Z}_c = \frac{1}{\Delta}\begin{bmatrix} G_0 + Y(\omega_1) & 0 & -G_{-2} \\ 0 & \dfrac{\Delta}{G_0 + Y(\omega_0)} & 0 \\ -G_2 & 0 & G_0 + Y^*(\omega_{-1}) \end{bmatrix} \tag{8.58}$$

where the determinant, Δ, is

$$\Delta = G_0^2 + G_0 Y(\omega_1) + G_0 Y^*(\omega_{-1}) + Y(\omega_1) Y^*(\omega_{-1}) + G_2 G_{-2} \tag{8.59}$$

For purposes of the example, let us assume that the noise source, i_n in Figure 8.12, has a $1/f$ spectrum and is modulated by the voltage across it. Then,

$$\overline{|i_n|^2} = \frac{K_f V_1^2 (1 + \cos(\omega t))^2}{f} \tag{8.60}$$

where K_f is a constant. From Section 6.2.3, we obtain the correlation matrix,

$$\mathbf{C}_i = \frac{K_f V_1^2}{4f}\begin{bmatrix} 1 & 2 & 1 \\ 2 & 4 & 2 \\ 1 & 2 & 1 \end{bmatrix} \tag{8.61}$$

Substituting (8.58) and (8.61) into (8.57) to obtain the voltage correlation matrix results in a long algebraic expression. Repeating it here serves little purpose, but it is worthwhile to examine a noise component, say, the upper sideband. It is

$$\overline{|v_{usb}|^2} = \frac{K_f V_1^2}{4f|\Delta|^2}[(G_0 + Y(\omega_1) - G_{-2})(G_0 + Y^*(\omega_1)) + (-G_0 - Y(\omega_1) + G_{-2})G_{-2}] \tag{8.62}$$

Finally, the output noise power, N_{ssb}, is

$$N_{ssb} = \overline{|v_{usb}|^2} G_L \tag{8.63}$$

and, of course, the sinusoidal output power, P_o, is

$$P_o = \frac{V_1^2}{2} G_L \tag{8.64}$$

Dividing these gives the SSB noise-to-carrier ratio, which we plot in Figure 8.15. This plot, which includes both AM and phase noise, agrees well with calculations by a circuit simulator. It also shows the expected 30-dB/decade slope.

However, note what happens when we modify a term slightly. We change G_2 from 0.005 to 0.005000001. The result is shown in Figure 8.16.

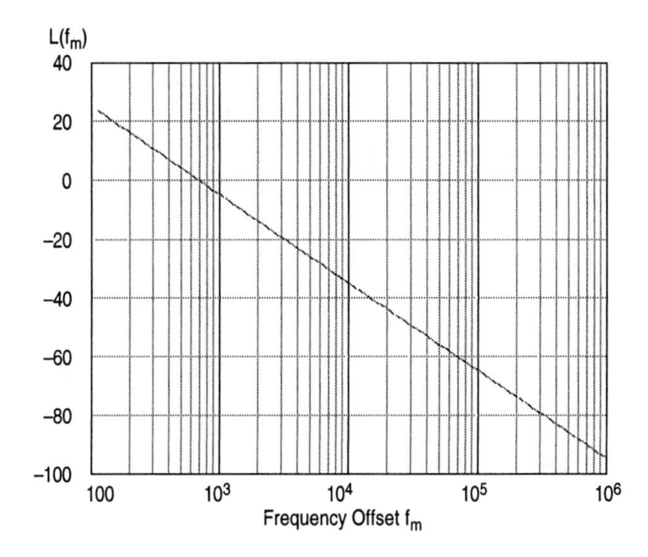

Figure 8.15 Calculated carrier-to-noise ratio of the Van der Pol oscillator.

The close-in phase noise has changed dramatically, as a result of a miniscule error in one harmonic term, although the results above 10 kHz remain unchanged. This illustrates why the unmodified conversion-noise analysis does not work well for oscillator analysis.

We can find out a little more about the oscillator's noise characteristics by examining the conversion matrix in the absence of the resonator. It has the form,

$$\mathbf{G} = (G - G_L)\begin{bmatrix} 1 & 0 & 1 \\ 0 & 1 & 0 \\ 1 & 0 & 1 \end{bmatrix} \tag{8.65}$$

which is obviously singular. Recall that, in a negative-resistance oscillator, the positive load conductance and negative device conductance sum to zero. In this case, they combine to form a singular admittance matrix, which is the matrix analog. It implies that the small-signal device admittance, at the fundamental frequency, has no real part. Therefore, when the combination is excited by a noise-current source, the noise voltage is determined entirely by the resonator's susceptance, which has a $1 / \omega^2$ magnitude.

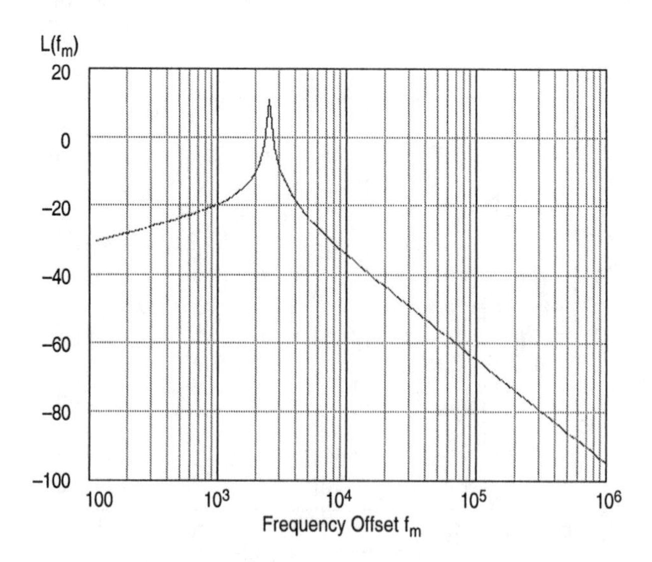

Figure 8.16 Carrier-to-noise ratio of the oscillator showing dramatic changes resulting from only a slight variation of G_2.

One might expect the average value of the junction conductance to cancel the load conductance. However, this is clearly not so. In this case, the average device conductance is $G - 2G_L = -0.005$ S, while the load is 0.010 S. If one were to assume that the device, for noise analysis, could simply be characterized by its average conductance, the total conductance would have a real part, and the expected 30-dB/decade slope would not be observed; in fact, one would observe the kind of characteristic that is often used to illustrate the inadequacy of conversion-noise analysis (see, e.g., [8.9]). Paradoxically, the singularity of the matrix, troublesome as it is, is essential for proper noise analysis.

This implies, in turn, that it is essential to use the entire conversion matrix for noise analysis. In the case of the Van der Pol oscillator, the matrix is small and finite, so use of the entire matrix is not difficult. In a larger problem, however, in which the nonlinearities are stronger and the voltage has, at least theoretically, an infinite number of harmonics, it may not be easy to determine where to truncate the series, and the noise analysis might depend strongly on the number of Fourier terms retained.

Finally, it is easy to see why the noise has the 30-dB/decade slope evident in Figure 8.15. The small-signal conductance is, as we have stated, zero, so only the resonator susceptance remains. It is excited by the upper-sideband component of the upconverted $1/f$ noise. For small deviations from the resonant frequency, $\Delta\omega$, the parallel LC resonator's reactance, is easily shown to be

$$X(\Delta\omega) = \frac{\omega_r}{\Delta\omega} \frac{1}{(2\omega_r C)} \tag{8.66}$$

where ω_r is the resonant frequency. Then,

$$\overline{|v_{usb}|^2} \approx \overline{|i_{usb}|^2} |X(\Delta\omega)|^2 \tag{8.67}$$

where i_{usb} is the upper-sideband component of i_n, from (8.60). The mean-square value of the latter quantity varies as $1/\Delta\omega$, while, from (8.66), $|X(\omega)|^2$ varies as $1/\Delta\omega^2$, so $\overline{|v_{usb}|^2}$ varies as $1/\Delta\omega^3$, creating a 30-dB/decade slope. This slope changes to 20 dB/decade when the spectral density of the $1/f$ noise drops below the white thermal noise of the resistors (which we did not include in the analysis). At even greater offsets, (8.66) no longer is valid, and the noise spectrum eventually levels out to broadband, white noise.

8.4 OPTIMIZATION OF LOW-NOISE OSCILLATORS

8.4.1 Devices

As the dominant noise source in oscillator phase noise is $1/f$ noise, it should be clear that devices exhibiting lower levels of such noise should have lower phase noise. Bipolar devices, both heterojunction and homojunction, have significantly lower $1/f$ noise than FETs, and thus are preferred. Bipolars, however, are more limited in frequency than microwave FETs, which can be used as oscillators well into millimeter wavelengths. At very high frequencies, FETs may therefore be the only option. The characteristics of these devices, as they apply to oscillators, have been discussed in detail in Section 8.3.2.

The general expression for $1/f$ noise, (2.54), shows that $1/f$ noise is proportional to a power of the collector or drain current. This implies that minimizing the collector or drain current, for a specified output power, should minimize the phase noise. Although $1/f$ noise dominates, it is possible, in some cases, for other noise sources to affect the oscillator's phase noise. For example, it is possible for the device to generate high levels of noise if avalanche breakdown occurs. Ordinary high-frequency noise of the transistor is rarely a dominant contributor to phase noise.

8.4.2 Resonator Q

Maximizing the Q of the oscillator's resonant circuit is the most effective way to minimize phase noise. We noted, however, in Section 8.3.1, that the Q is difficult to define precisely. For example, it is difficult to apply the concept of a loaded Q, as used in Leeson's model, to an oscillator that does not have an easily defined feedback path. Even in that case, it still may be impossible.

In all cases, however, the critical quantity is the loss in the frequency-determining parts of the circuit. Even relatively small losses can increase the noise level dramatically.

In feedback oscillators designed according to the approach in Section 8.1.5.2, maximizing the Q corresponds to maximizing $d\phi/df$, where ϕ is the loop phase shift of the amplifier-resonator combination, in the vicinity of resonance. This maximization can be accomplished by coupling the resonator weakly to the circuit. As the coupling is made weaker, however, the Q is limited by the resonators's unloaded Q, and the increased loss in the feedback circuit reduces the loop gain, which must remain comfortably above unity to allow oscillation.

In oscillators using dielectric or metal cavity resonators, the same principles apply. Circuit losses must be minimized, of course, and coupling to the resonator made relatively weak, consistent with the need to achieve oscillation. Dielectric-resonator oscillators (DROs) usually employ a resonator coupled to a microstrip line. The coupling to a dielectric resonator is a strong function of its distance from the line and, when critically coupled, the structure is equivalent to a high-Q, parallel LC resonator in series with the line. Coupling the resonator more strongly to the line reduces its Q, while weaker coupling can prevent oscillation entirely.

Many integrated-circuit realizations use planar spiral inductors in their resonators. Although small, planar inductors are notoriously lossy, and their low Qs generally result in high noise levels. This is especially true of inductors in CMOS and other silicon technologies, as they experience loss in the substrate as well as in their metal. Considerable research has been devoted to the improvement of the Qs of such inductors.

Voltage-controlled oscillators can be realized in a number of ways. The most common is to create a negative-resistance oscillator resonated by a varactor diode. Varactors are usually realized as *pn* junction devices, although, in some cases (e.g., millimeter wavelengths or in monolithic processes, where *pn* junctions cannot be fabricated), Schottky-barrier structures are used instead. The diode's Q at some frequency ω, Q_d, is defined as

$$Q_d = \frac{1}{\omega R_s C_j(V_b)} \tag{8.68}$$

where R_s is the series resistance, C_j is the junction capacitance, and V_b is some standard bias voltage. One standard value of V_b is –6.0V, a relatively high reverse voltage, which guarantees larger values of Q_d. This practice provides clear advantages in marketing the devices.

The diode's series resistance is the dominant loss mechanism in varactor-tuned VCOs. In VCOs using Schottky varactors, rectification in the diode, which results in conduction loss, can also degrade phase noise. In either case, Q_d decreases as tuning voltage increases, per (8.68), so the oscillator's phase noise varies with tuning voltage as well.

There is, in general, a trade-off between tuning range in VCOs and varactor Q. Wide tuning range requires that the varactor be strongly coupled to the circuit, reducing the loaded Q. Some types of diodes offer wide tuning range at the cost of lower Q; for example, hyperabrupt varactors, which have a wide capacitance range, also exhibit greater series resistance than simple *pn* junction varactors.

8.4.3 Other Methods

Other methods have been proposed for minimizing phase noise in oscillators. We discuss a few of them briefly.

8.4.3.1 Feedback

Because $1/f$ noise is a baseband phenomenon, it should be possible for low-frequency feedback to minimize $1/f$ noise without affecting the oscillation process. Occasionally, surprisingly large improvements are claimed for such methods. In [8.10], for example, negative feedback is used in a bipolar transistor oscillator to reduce the phase noise by 40 dB compared to the same circuit without feedback.

8.4.3.2 Coupled Oscillators

Another method for reducing phase noise involves coupling a number of synchronized oscillators. It can be shown that coupling N oscillators can reduce the phase noise by a factor of $1/N$. This technique can be useful in certain kinds of power-combining or beam-scanning applications, which inherently require a large number of oscillators [8.11].

8.4.3.3 Analytical Methods

Many researchers have described design methods for optimizing oscillators. Although there are far too many in existence to provide a comprehensive list, we list a few examples in the references [8.12–8.16]. It is interesting to note the lack of uniformity in the methods; each takes a significantly different approach from the others. For example, [8.12] is a time-domain approach, in which an impulse response for the phase and amplitude of the oscillation are first described. This leads to the definition of an *impulse sensitivity function* (ISF), whose average is critical to the noise characteristics. The authors then show that this function is minimized by certain symmetries in the oscillator's waveform, and that phase noise can be minimized by creating those symmetries. The ISF is used in [8.13] but, because it is periodic, the ISF is Fourier transformed and used in a harmonic-balance method.

In [8.14], the oscillator is optimized by building a kind of optimization functionality into the basic Newton iterations of the harmonic-balance process. In somewhat the same idea, [8.15] treats the oscillator design as a numerical optimal control problem. Finally [8.16] notes that the oscillator may limit on current, voltage, or both, and optimizes the oscillator on the basis of those limiting methods.

The lack of commonality in approaches to a new field of research is not unusual. In the beginning, researchers grope in the dark for approaches to the subject, but eventually converge on a consensus as to the best methods. It is surprising, in the case of oscillators, to see this process continuing more than a decade. It is evidence of the immaturity of the field, and the need for further research to unify and generalize it.

References

[8.1] K. Kurokawa, "Some Basic Characteristics of Broadband Negative Resistance Oscillator Circuits," *Bell Sys. Tech. J.*, Vol. 48, p. 1937, 1969.

[8.2] R. Rhea, *Oscillator Design and Computer Simulation*, 2nd ed., New York: McGraw-Hill, 1995.

[8.3] W. Wagner, "Oscillator Design by Device-Line Measurements," *Microwave J.*, Vol. 22, No. 2, p. 43, February 1979.

[8.4] E. Ngoya, J. Rousset, and D. Argollo, "Rigorous RF and Microwave Oscillator Phase Noise Calculation by Envelope Transient Techniques," *IEEE International Microwave Symposium Digest,* 2000.

[8.5] E. Ongareau, F. M. Ghannouchi, and R. Bosisio, "Harmonic Device Line Simulation of Negative Resistance Microwave MESFET Oscillators," *Microwave and Optical Technology Letters*, Vol. 3, p. 317, 1990.

[8.6] S. A. Maas, *Nonlinear RF and Microwave Circuits*, Norwood, MA: Artech House, 2003.

[8.7] D. B. Leeson, "A Simple Model of Feedback Oscillator Noise Spectrum," *Proc. IEEE*, Vol. 54, p. 329, February 1966.

[8.8] D. Scherer, "Today's Lesson—Learn About Low-Noise Design," *Microwaves*, p. 116, April 1979.

[8.9] V. Rizzoli, F. Mastri, and D. Masotti, "A General-Purpose Harmonic-Balance Approach to the Computation of Near-Carrier Noise in Free-Running Microwave Oscillators," *IEEE International Microwave Symposium Digest,* p. 309, 1993.

[8.10] U. Rhode and F. Hagemeyer, "Feedback Technique Improves Oscillator Noise," *Microwaves and RF,* p. 61, November 1988.

[8.11] H.-C. Chang et al., "Phase Noise in Coupled Oscillators: Theory and Experiment," *IEEE Trans. Microwave Theory Tech.*, Vol. 45, p. 604, 1997.

[8.12] A. Hajimiri and T. H. Lee, "A General Theory of Phase Noise in Electrical Oscillators," *IEEE J. Solid State Circuits*, Vol. 2, p. 179, 1998.

[8.13] S. Ver Hoeye, A. Suárez, and J. Portilla, "Techniques for Oscillator Nonlinear Optimization and Phase-Noise Analysis Using Commercial Harmonic-Balance Software," *2000 IEEE MTT-S International Microwave Symposium Digest*, 2000.

[8.14] V. Rizzoli et al., "Harmonic-Balance Optimization of Microwave Oscillators for Electrical Performance, Steady-State Stability, and Near-Carrier Phase Noise," *1994 IEEE MTT-S International Microwave Symposium Digest 1994.*

[8.15] V. Güngerich et al., "Phase Noise Reduction of Microwave Oscillators by Optimization of the Dynamic Behaviour," *1994 IEEE MTT-S International Microwave Symposium Digest, 1994.*

[8.16] W. Anzill et al., "Phase Noise Minimization of Microwave Oscillators by Optimal Design," *1995 IEEE MTT-S International Microwave Symposium Digest*, 1995.

Chapter 9

Mixers and Frequency Multipliers

From the MIT Radiation Laboratory days of the 1940s until the successful development of high-frequency gallium arsenide MESFETs in the 1980s, low-noise receiver front ends used mixers. Considerable research on low-noise mixers was carried out in that period, culminating in the classic 1976 paper by Held and Kerr [9.1] describing the first practical noise model of a millimeter-wave mixer. Today, mixer noise modeling, although generalized, is performed in much the same manner as described by Held and Kerr.

Although the theoretical aspects of mixer noise modeling are now well understood, the subject is still complicated by a number of practical dilemmas. Chief among these is the handling of diode or active-device terminations at mixing frequencies beyond those of the RF input and IF output. The problem is further complicated by a number of popular publications that are poorly thought out or, occasionally, erroneous. To minimize the visibility of such publications, we will not reference them here; instead we present the correct theory, and hope that the incorrect simply disappears.

9.1 ESSENTIAL MIXER THEORY

A mixer is fundamentally a multiplier, which multiplies the signal frequency, usually called the *RF*, by a local oscillator waveform, the *LO*, ideally sinusoidal, to produce an output at the intermediate frequency, or *IF*.[1] The

1. Sometimes the term *RF* is applied to the high-frequency signal and *IF* to the low-frequency, regardless of their status as input or output. That is, the input signal in an up-converter is sometimes called the *IF* and the output, the *RF*. In this book we shall avoid this complication and standardize the terms as described here.

signal at the IF is a replica of the RF, shifted to a lower or higher frequency, with (again, ideally) no loss of information.

The mixing process requires a time-varying parameter, which is usually a conductance or a transconductance. Time-varying reactances also can produce mixing, but they have significant disadvantages and are rarely used today.[2] A mixer using a time-varying conductance is called a *resistive mixer*; the conductance usually is realized by pumping a nonlinear resistance (or controlled current source) by the LO. A mixer in which mixing occurs by virtue of time-varying transconductance, usually realized by a FET or bipolar device, is called an *active mixer.*

9.1.1 Resistive Mixers

Schottky-barrier diodes are most frequently used in resistive mixers. A diode has the I/V characteristic

$$I(V) = I_0\left[\exp\left(\frac{qV}{\eta KT}\right) - 1\right] \tag{9.1}$$

Its conductance waveform, $g(t)$, is

$$g(t) = \left.\frac{dI}{dV}\right|_{V = V_{\mathrm{LO}}(t)} = \frac{qI_0}{\eta KT}\exp\left(\frac{qV_{\mathrm{LO}}(t)}{\eta KT}\right) \approx \frac{q}{\eta KT}I(V_{\mathrm{LO}}(t)) \tag{9.2}$$

where $V_{\mathrm{LO}}(t)$ is the junction-voltage waveform generated by the LO signal. $V_{\mathrm{LO}}(t)$ is invariably nonsinusoidal, as the LO signal is distorted by the diode's nonlinearity. Mixing occurs through the process

$$i(t) = g(t)v(t) \tag{9.3}$$

where $i(t)$ and $v(t)$ are the small-signal voltage and current in the diode. These consist of the RF and IF components, as well as a large number of mixing frequencies, in which the latter mix with harmonics of the LO. In general, the mixing frequencies are given by

2. A set of equations called the Manley-Rowe relations shows that the maximum conversion efficiency of a positive-resistance reactive mixer is equal to the ratio of the output to input frequency. Thus, a 12-GHz mixer with a 60-MHz IF must have *at least* 23-dB loss! On the other hand, an upconverter can have gain. In the past, reactive upconverters have been used as low-noise amplifiers.

$$\omega_k = k\omega_p \pm \omega_0 \tag{9.4}$$

where ω_p is the LO frequency, ω_0 is the lowest-frequency mixing product (the IF in a downconverter), and k is an integer.

The large-signal voltage and current waveforms in the diode can be determined by either time-domain or harmonic-balance methods, although the latter is most often employed. The mixer is analyzed first under LO excitation alone, then $g(t)$ is determined from (9.3), and a time-varying analysis is performed. This is a direct application of conversion-matrix analysis, described in Sections 6.1.3 and 6.1.4. The analysis in those sections shows that the mixer can be viewed as a kind of multiport, in which the ports are the voltages and currents at the individual mixing frequencies, instead of physically separate ports. Indeed, at least in concept, physically separate ports could be created; ideal filters could separate the mixing currents and voltages into individual ports. Figure 9.1 illustrates this situation.

From the figure and the development in Sections 6.1.3 and 6.1.4, we can see that there is no difference, mathematically, between the formulation of the time-varying mixer, described by conversion matrices, and a conventional multiport described by an admittance (or other) matrix. We know that the transfer function between any pair of ports in a multiport network depends on the terminations of the other ports. In the same way, the conver-

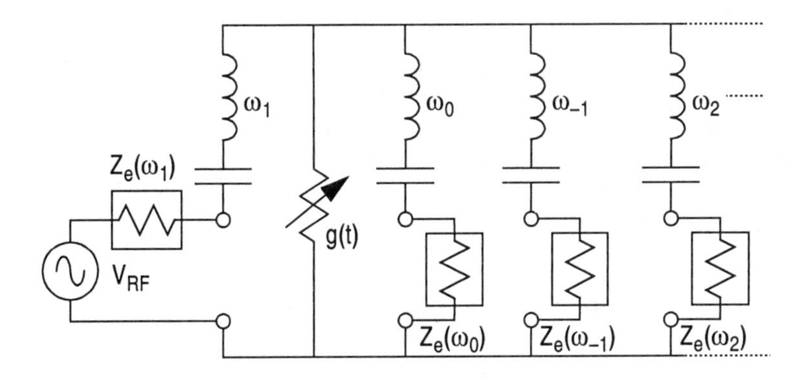

Figure 9.1 A mixer with lossless matching circuits can be viewed as a multiport. Ideal resonators are used to separate the mixing products, and each port—which corresponds to a mixing frequency—is terminated by an embedding impedance. In this illustration, we have assumed that the RF excitation is ω_1 (9.4) and the IF is ω_0.

sion efficiency and, in fact, all aspects of a mixer's operation, depend on the terminations of the diode at the unused mixing frequencies.

We call the impedances that terminate the diode at the mixing frequencies (9.4) the set of *embedding impedances* seen by the diode. All aspects of the performance of any mixer depend only on the mixing element, the set of embedding impedances at the mixing frequencies given by (9.4), and the $g(t)$ waveform. The latter is determined by the LO level and the embedding impedances at the LO harmonics.

In practical mixers, the LO and RF are applied through appropriate filter/matching circuits. These are shown, for a resistive mixer, in Figure 9.2. The embedding impedances in that case consist of the port impedances of the filter/matching networks. For best conversion efficiency, we want no power dissipated in the networks, so the embedding impedances at frequencies other than the RF and IF ideally should be reactive. The nature of the reactances can have a strong effect, however, on the performance of the mixer. In general, short-circuit terminations provide well-behaved operation and reasonable input and output impedances. Other terminating impedances, especially at the image and sum frequencies, can have a dramatic effect on the conversion loss and port impedances; for example, open-circuit terminations at these frequencies can create a high RF input impedance. In concert with the diode's pumped junction capacitance, they can even create parametric oscillation.

Optimizing the performance of a mixer becomes an exercise in terminating the ports/mixing frequencies ideally. Much of the early research in mixers was devoted to understanding the effects of those terminations. In practice, however, we generally cannot terminate a large number of frequency components optimally; it simply is not practically possible. Instead, we attempt to identify the components that have the strongest effect on conversion performance and noise, and we terminate them as best we can. In diode mixers, the most important components (aside from the RF and IF), are the LO harmonics, and the lower-order mixing products.

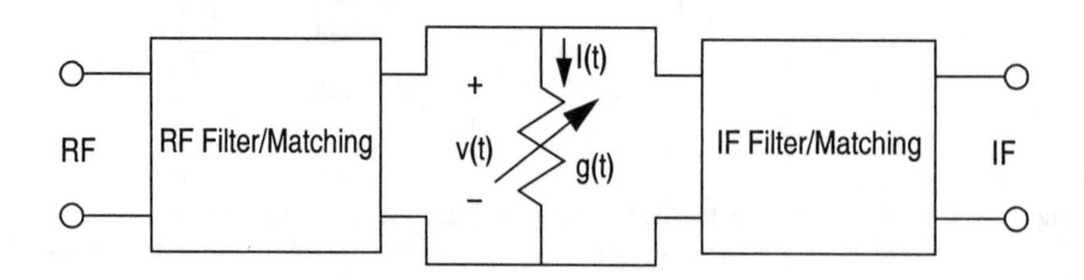

Figure 9.2 Mixer using a time-varying conductance.

9.1.2 FET Resistive Mixers

Another type of resistive mixer is called a *FET resistive mixer* [9.2, 9.3]. Such a mixer uses a FET's resistive channel as the mixing element. As shown in Figure 9.3, the LO is applied to the FET's gate, the RF to the drain, and the IF is extracted from the drain. The gate is biased near pinch-off, but the dc drain bias is set at zero, so the device is not active. Instead, the LO varies the FET's channel resistance, creating a pulsed conductance waveform that is similar to the conductance waveform of a diode.

Because the conductance waveform is similar to that of a diode mixer, the FET resistive mixer's conversion performance is similar. The conversion loss is roughly the same, approximately 6 to 8 dB in the microwave region, and input and output impedances are moderate and easy to match. Because the FET channel is more linear than a diode junction, however, FET resistive mixers have much lower distortion and higher saturation levels than diode mixers.

A disadvantage of the FET resistive mixer is the FET's large parasitic capacitances, relative to a diode. When the drain is zero-biased, the FET's gate-to-channel capacitance is roughly evenly divided between the gate-to-drain capacitance and gate-to-source. As a result, significant coupling of the LO to the drain, which could generate noise and distortion, is possible unless special care is taken in designing the terminating impedances at the drain. As with an active FET mixer, in the FET resistive mixer it is essential to short-circuit the drain at the LO frequency.

The noise of a FET resistive mixer is entirely the thermal noise of the channel, with small contributions from the parasitic resistances. The channel carries no dc or large-signal current, so it generates no $1/f$ noise, and

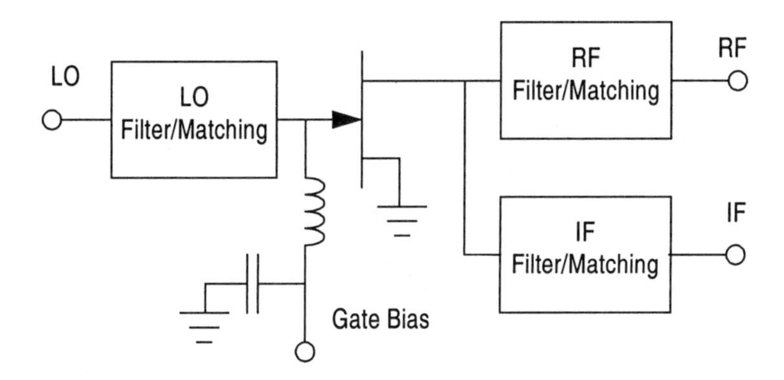

Figure 9.3 FET resistive mixer.

as with all FETs, it generates no shot noise. The absence of low-frequency noise makes the FET resistive mixer attractive for receivers that convert the received signal directly to baseband.

9.1.3 Active Mixers

Mixers can also be realized by pumping an active device, usually a FET or bipolar device, to vary its transconductance. The circuit of an active mixer is shown in Figure 9.4, including its noise sources. Mixing occurs by virtue of a time-varying transconductance, created by applying the LO voltage to the gate. Active mixers include a number of other nonlinearities, which have some effect on the mixing action. The gate-to-source capacitance of a FET, for example, and its drain-to-source resistance are minor contributors to mixing.

As with the resistive mixer of Figure 9.1, the embedding impedances are those of the input and output matching circuits. The requirements for the embedding impedances in active mixers are, as one might expect, significantly different from those for diode mixers. In active mixers, the drain terminations at the LO harmonics are critical; the fundamental frequency and, if possible, an additional harmonic or two should be short-circuited. The gate terminations are only slightly less critical; shorting the gate at the IF is also important. We discuss the optimization of these terminations in detail in Section 9.4.2.

The noise properties of active mixers are also very different from resistive ones. In diode mixers, the noise is predominantly shot noise. In active FET mixers, high-field diffusion noise in the drain dominates, and thermal noise in the gate and source resistances, while lesser, are important noise sources. In bipolar devices, shot noise in the collector and thermal noise in

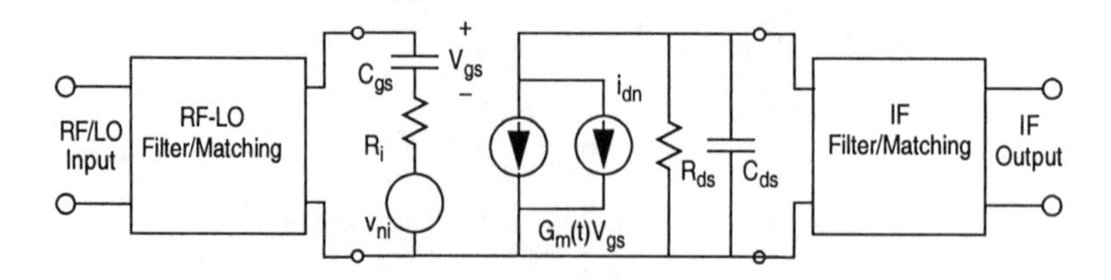

Figure 9.4 Active transconductance mixer. As with an amplifier, the noise sources consist of thermal noise associated with the input resistance and an output noise source associated with the channel.

the base resistance are the primary noise sources. The behavior of the noise in an active circuit is naturally very different from a passive one.

9.2 NOISE IN DIODE MIXERS

9.2.1 Noise Characteristics

Noise in diode mixers comes primarily from shot noise in the diode junction. Thermal noise from the series resistance of the diodes, as well as circuit losses in the mixer, are additional, but lesser, contributors to the mixer's noise temperature. The series resistance not only generates thermal noise, but, more significantly, increases the mixer's conversion loss. This indirect effect of the resistance—increased loss—has a greater impact on noise temperature than the direct effect of its thermal noise.

Early analyses of diode mixers indicated that a mixer's noise should be well described by the attenuator noise model (Section 3.2.1) with a diode temperature of $\eta T / 2$, where η is the slope parameter of the diode (9.1) and T is its physical temperature [9.4]. Measurements of practical mixers, however, indicated that the diode's effective noise temperature was consistently double this value, occasionally even higher. The discrepancy was resolved by mixer noise models that included all the correlation properties of the modulated shot noise [9.1]. This understanding forms the basis of the nonlinear noise analysis implemented in modern circuit simulators.

Because the mixer's embedding impedances affect both the correlation properties of its shot noise and its conversion efficiency, a mixer's noise performance depends strongly on the terminations of the diode at all the mixing frequencies given by (9.4). These terminations must be optimized to achieve low-noise mixing.

Diodes also generate $1/f$ noise. This low-frequency noise is a concern only when the IF extends close to baseband. That is the case in many kinds of receivers (e.g., FMCW radars) and mixers used as phase detectors. It is also a concern in diode detectors. Low-frequency noise levels are less predictable than high-frequency, as they depend more strongly on the physical characteristics of the diode itself. Some diodes have higher levels of low-frequency noise than others; for example, many zero-bias detector diodes use Schottky barriers on p material, instead of the usual n material, in part because the noise level is lower.

9.2.2 Optimization

9.2.2.1 General Considerations

Not long ago, low-noise amplifiers above a few gigahertz did not exist, so low-noise receivers used diode-mixer front ends. In those days, the mixer's noise temperature was dominant in establishing the receiver's noise temperature, so its minimization was extremely important. Today, however, microwave receivers use low-noise amplifiers (LNAs) to establish the noise temperature, and the noise temperature of the mixer is less critical.

Even so, the mixer's noise temperature is rarely negligible, because the LNA gain ahead of the mixer must be limited. Because mixers use nonlinear elements, and the small devices used in LNAs cannot handle large signals very well, distortion in the mixer and LNA can be a more serious problem in such receivers than in the older receivers using mixer front ends. Too much gain can exacerbate nonlinear distortion in both the mixer and LNA.

Thus, in the design of a receiver, it is necessary make the LNA gain high enough to achieve a low receiver noise temperature, yet low enough to keep distortion levels under control. This can become a delicate balancing act, which is made easier when the mixer's noise temperature is reasonably low. That noise temperature is not as critical in receivers using an LNA front end as in a diode-mixer front end, but it is clearly a quantity of concern.

9.2.2.2 Conversion Loss

The designer has relatively few degrees of freedom in optimizing the noise temperature of a diode mixer. The most effective method for minimizing a mixer's noise temperature is to minimize its conversion loss. This requires careful design, especially attention to RF input and IF output matching and to the embedding impedances at LO harmonics and the sum and image frequencies. The diode parasitics must also be minimized, for example, by the use of GaAs instead of silicon diodes.

It is generally not too difficult to achieve conversion loss, in narrow- or moderate-bandwidth mixers, below 6 dB in the microwave range. Somewhat higher losses should be expected at millimeter wavelengths. The conversion loss of broadband mixers is generally somewhat higher, as many more mixing products may be terminated in the resistive source and load impedances.

9.2.2.3 Image and Sum Terminations

The termination of low-order mixing products, especially the image and sum frequencies, can strongly affect the conversion loss of the mixer. It also can affect distortion properties and the diode impedance at the RF and IF frequencies. Proper termination of these products can do much to minimize loss and thus noise temperature. In the past, an adjustable image or sum termination was used specifically to optimize the conversion loss in diode-mixer front ends. Such *image-enhanced* or *sum-enhanced* mixers occasionally achieved conversion losses below 3 dB. Unfortunately, the low conversion loss was often accompanied by a relatively high effective diode temperature, giving only a modest improvement in the mixer's noise temperature.

Frequently a filter is used ahead of the mixer to eliminate its image response. Then, the mixer is image-enhanced by default. The image termination should then be adjusted to optimize the mixer conversion loss or at least to make sure that the performance is not degraded. The adjustment can be made by varying the distance between the mixer and the filter.

The image termination may affect distortion characteristics as well as conversion loss and noise temperature. It has been shown that the point of minimum conversion loss actually can have relatively high levels of distortion [9.5]. Short-circuiting the diode at the image frequency minimizes distortion and still provides good (although perhaps not minimum) conversion loss.

9.2.2.4 Temperature

Cooling a diode mixer decreases its noise temperature. Decreasing the thermal noise in the diodes' series resistance and in circuit losses is a relatively minor cause of the noise-temperature reduction. The primary effect is the strengthening of the diode's nonlinearity as temperature decreases, which allows minimum conversion loss to be achieved at lower LO power. This results in lower junction current, and, finally, less shot noise in the diode.

In the past, cooling—often to cryogenic temperatures—was commonly used to achieve low noise temperatures in diode-mixer front ends. Reduction in mixer noise temperature by a factor of three to four can be achieved with cooling to ~15K. Today, this practice is largely obsolete, at least for mixers in the microwave and millimeter-wave regions. Cooling still is used occasionally to minimize the noise temperature of low-noise amplifiers in radio-astronomy receiver front ends, and in submillimeter-wave mixer receivers. In cooled-amplifier front ends, the mixer can be cooled with little additional difficulty.

9.2.2.5 Other Noise Sources

Especially with silicon diodes, which have relatively low breakdown voltages, it is important to avoid driving the diodes, with the LO power, into reverse breakdown. If breakdown occurs, excess noise is generated and the conversion loss increases. GaAs diodes have higher breakdown voltages than silicon, so the former should be used in applications where breakdown might occur. In doubly balanced ring or star mixers, the diodes are effectively antiparallel, so each one limits the reverse voltage of another. This makes reverse breakdown virtually impossible.

Another source of noise is parametric instability, which can occur, albeit rarely, in mixers. Such instability is caused by the pumped junction capacitance and is similar to oscillation in varactor frequency multipliers. In mixers, however, the junction capacitance is loaded by the average value of the junction conductance, which is usually adequate to prevent such oscillation. If parametric instability occurs, noise levels can become quite high. Such instability introduces so many problems that, should it occur, the noise level of the mixer might be the smallest of the problems that a designer must deal with.

9.3 NOISE IN FET RESISTIVE MIXERS

9.3.1 Noise Characteristics

Although the conversion performance of a FET resistive mixer is similar to that of a diode mixer, the noise generating mechanism is very different. The noise in a properly designed FET resistive mixer is entirely the thermal noise of the channel and the drain and source parasitic resistances. Unless the gate is driven into conduction, which is clearly undesirable, the FET generates no shot noise or $1/f$ noise.

The thermal noise of the channel comes from a modulated thermal noise source. The treatment of such sources is described in Section 6.2.2. From fundamental physical considerations, the available noise power must be the same as from a static resistance; thus, a FET resistive mixer literally obeys the attenuator noise model, with an effective device temperature equal to its physical temperature.

9.3.2 Optimization

As with any resistive mixer, the noise temperature of a FET resistive mixer is minimized when the conversion loss is minimized. Therefore, the first effort in optimizing the noise performance of a FET resistive mixer is to minimize its conversion loss. This must be accomplished, however, with appropriate drain terminations and without driving the FET's gate into conduction or breakdown. Such operation results in the generation of shot noise, which significantly increases the noise temperature of the mixer.

The noise performance of a FET resistive mixer depends strongly on its terminations. The requirement of keeping the drain voltage at zero implies that the drain must be short-circuited at the LO frequency and, as with other types of FET mixers, at as many harmonics as is practically possible. This requirement may conflict with the need to match the drain at the RF, especially when the RF frequency is close to the LO; in that case, the use of a singly or doubly balanced mixer can provide the proper drain terminations. Short-circuiting the gate at the RF and IF frequencies is less important than in active mixers, but still may be helpful.

It is essential to prevent the gate from being driven into conduction or breakdown by the LO. At the same time, minimizing distortion in such mixers requires high LO level. Thus, avoiding gate conduction may be difficult in practical systems, where the LO level may drift with temperature or vary with frequency. In such cases, a large resistor (a few thousand ohms) can be placed in series with the gate bias source. This resistor provides negative bias feedback, making the gate bias more negative when dc current is generated.

9.4 NOISE IN ACTIVE MIXERS

9.4.1 Noise Characteristics

In Chapter 7 we saw that the noise performance of a low-noise amplifier depended on the interplay between the effects of the noise sources at the input and output of the active device. We noted that the effect of thermal noise at the input of the device could be minimized by mismatching the input. Increasing the source impedance increased the source voltage, for a given available power, making it large relative to the input thermal noise. Unfortunately, this mismatch decreased the gain of the amplifier, weakening the signal at the output and reducing the signal relative to the output noise.

The importance of one phenomenon or the other depends on the gain of the device, which varies with frequency. We saw, in amplifiers, that input mismatching has a greater effect at low frequencies, where the device gain is inherently high. However, as frequency increases, and gain decreases, the optimum source impedance moves closer to a conjugate match. Figure

7.4 illustrates this behavior, for a specific device, as a function of frequency. It is clear that the optimum source impedance approaches an open circuit at low frequencies. As frequency increases, the source impedance approaches a conjugate match.

The same general considerations apply to active mixers, but the conclusions must be modified in view of the mixer's lower gain. An active mixer is generally operated as a transconductance device, and, as such, is not very different in its noise behavior from an amplifier. In a transconductance mixer, the local oscillator voltage is applied to the gate (we assume, for the moment, that our device is a FET) causing the transconductance of the device to vary with time. The RF is applied to the gate as well, and the IF is filtered from the drain. Then, we have

$$i_{d,\,\mathrm{IF}}(t) = g_m(t)\, v_{g,\,\mathrm{RF}}(t) \tag{9.5}$$

where $g_m(t)$ is the time-varying transconductance, $v_{g,\,\mathrm{RF}}(t)$ is the gate RF voltage, and $i_{d,\,\mathrm{IF}}(t)$ is the IF drain current. The transconductance waveform is generally a train of approximately half-sinusoidal pulses and has the Fourier series,

$$g_m(t) = \sum_{q=0}^{\infty} G_{m,\,q} \cos(q\omega_p t + \phi_q) \tag{9.6}$$

with

$$v_{g,\,\mathrm{RF}}(t) = V_R \cos(\omega_R t) \tag{9.7}$$

Of course, the gate voltage includes other frequency components. Those are generally small, however, so we can ignore them for present purposes. Also ignoring the phase terms in (9.6), we obtain

$$i_{d,\mathrm{IF}}(t) = \ldots + \frac{V_R G_{m,\,1}}{2}[\cos((\omega_p - \omega_R)t) + \cos((\omega_p + \omega_R)t)] + \ldots \tag{9.8}$$

indicating that the fundamental-frequency component of the transconductance waveform dominates in providing conversion to the IF. Furthermore, that component converts the RF equally to the sum frequency and the difference frequency, one of which is invariably rejected. In a sense, half of the $G_{m,\,1}$ component is unavoidably wasted.

Clearly, we need to maximize $G_{m,\,1}$ in any practical active mixer. This quantity is maximum when the device is biased near pinch-off and the LO voltage is great enough for the peak transconductance to reach the device's maximum. $G_{m,\,1}$ can approach, at best, half the peak transconductance of the device, which is invariably significantly less than the transconductance when used as an amplifier. This smaller value of transconductance, in concert with the inevitable factor of 1/2 seen in (9.8), means that the active mixer's gain is considerably lower than that of an amplifier using the same device.

At the same time, pumping the device with the LO varies the level of nonthermal noise sources in the output. These are high-field diffusion noise in FETs and shot noise in bipolar devices. The modulation of these sources creates correlations between the noise sidebands, which generally make the output noise greater than in an amplifier.

In essence, we have a component not unlike an amplifier, but with less gain and greater noise. As we saw in Chapter 7, in such a case the benefit of input mismatching to improve noise temperature is minimal or nonexistent. In fact, we find empirically that there is not much benefit, if any, to mismatching the input of an active microwave mixer for noise optimization. At high frequencies, a conjugate match is about as good as it gets. At lower frequencies (say, in the RF range), there may still be some advantage to mismatching the gate to improve noise temperature.

9.4.2 Optimization

An active mixer that is optimized in a general sense—for conversion performance and input and output match—is close to optimum, if not literally optimum, for noise. The emphasis on low-noise active mixer design is not so much on doing something special to assure low noise, but to avoid certain pitfalls that can increase the noise temperature of an otherwise well designed circuit. The following sections describe some of the issues that can arise in such designs.

9.4.2.1 LO and Mixing-Frequency Terminations

The best overall performance of an active mixer occurs when the device is terminated at the gate and drain by short circuits at frequencies other than the RF at the gate and IF at the drain. It is especially important to short-circuit the drain or collector at the LO fundamental frequency and harmonics. This guarantees that the drain remains at its dc bias voltage throughout the LO cycle, maximizing the transconductance variation and therefore its fundamental-frequency component. It also minimizes any tendency toward instability and reduces amplifier-mode gain. Appropriately terminating the drain at the LO harmonics is unquestionably the most important task in designing a low-noise active mixer.

Terminating all mixing products optimally is, of course, impossible, as the number is theoretically infinite. It is usually possible, however, to short-circuit the first few LO harmonics and RF at the drain and the IF at the gate. These are the most critical components, and it is usually practical to terminate them correctly. The termination often can be realized by filters or stubs at the drain and gate. The gate termination can be integrated with the bias circuit.

Balanced mixers are valuable in this regard, as certain kinds provide the optimum termination by virtue of their circuit configuration, so filters or stubs are not necessary. The termination properties depend on the type of balanced mixer; see [9.6] for further information on their properties.

In a transconductance mixer, it is generally best to conjugate match the RF, LO, and IF ports. Matching the RF port, as we noted in Section 9.4.1, minimizes the noise figure and maximizes the conversion efficiency; conjugate matching the LO simply minimizes LO power requirements. Therefore, if one must choose between the two, it is usually better to favor the RF match. The IF output impedance in a gate-pumped transconductance mixer is often very high, and it may be impractical to consider a conjugate match. In this case, it is necessary to terminate the drain in an appropriate low (~50Ω to 100Ω) IF impedance.

The short-circuit termination of unwanted mixing products is optimum in a practical sense. That is, it results in a well-behaved circuit having good, predictable performance. It is not clear that this is optimum in a theoretical sense, however; it may be possible to achieve better performance, in some respect, with other terminations. For example, it is likely that higher conversion efficiency might be obtained with different terminations, although possibly at the expense of stability. Usually, however, the short-circuit criterion provides a straightforward, practical basis for good design, and it rarely makes sense to deviate from it.

9.4.2.2 Amplifier-Mode Gain

Active devices are fundamentally amplifiers. Pumping one with an LO signal does not change that fact. Unless special effort is taken to prevent it, the FET or bipolar transistor used in a mixer can amplify IF signals applied to its input.

In Section 9.4.2.1 we noted that short-circuiting the device's gate at all frequencies except the RF was essential to achieving good performance. The elimination of amplifier-mode gain is one reason for this. If the gate is not shorted, any IF-frequency signals or noise leaking into the gate are simply amplified and transmitted to the output, effectively increasing the output noise temperature and therefore the mixer's effective input noise temperature. One source of such signals is IF-frequency noise in the bias circuit. This is a major cause of high noise figures in active mixers.

Amplifier-mode gain can also be a source of instability. Instability in active mixers is especially insidious, as it can be caused by amplifier-mode gain at any frequency, not just at the mixing frequencies. Again, the recommended short-circuit terminations can do much to minimize this problem.

9.5 SYSTEM CONSIDERATIONS

9.5.1 Spurious Responses

All mixers have responses at frequencies other than those desired. In general, a mixer produces an IF output at the frequencies

$$\omega_{IF} = m\omega_{RF} + n\omega_p \tag{9.9}$$

where ω_p is the LO frequency and m, n are integers. An (m, n) pair that creates an in-band IF signal is called *an (m, n) spurious response*. A spurious response can be especially troublesome if the RF frequency associated with it falls within the RF band; then, it cannot be eliminated by filtering. Noise signals are invariably too weak to generate RF harmonics, but they can enter the mixer by mixing with LO harmonics. Therefore, for noise analysis, we are especially concerned with $(\pm 1, n)$ responses; the worst of these responses is the image, $2\omega_p - \omega_{RF}$.

The existence of these responses creates a dilemma in using and characterizing a mixer. The dilemma is illustrated by Figure 9.5, in which a mixer is preceded by a low-noise amplifier. We consider, for the moment, only the RF and image responses. The mixer has an explicit pair of inputs, the image and RF; in fact, these are separate responses at a single input, but we treat them as a pair of inputs for illustrative purposes. The output noise temperature of the mixer (Section 3.2.1), $T_{m,L}$, is

$$T_{m,L} = T_n G_a G_{cr} + T_{SSB} G_{cr} + T_{a,L} G_{ci} \tag{9.10}$$

where G_a is the amplifier's RF gain and $T_{a,L}$ is its output noise temperature at the image frequency. The effective input noise temperature of the combination, T_s, is

$$T_s = T_n + \frac{T_{SSB}}{G_a} + \frac{T_{a,L}G_{ci}}{G_aG_{cr}} \tag{9.11}$$

The problem, in this case, is $T_{a,L}$. If we assume that $G_{ci} = G_{cr}$ and that the amplifier's gain and noise temperature are the same at the image frequency as the RF, $T_{a,L}/G_a = T_n$ and the system noise temperature is approximately doubled, compared to a mixer without an image response. However, the situation could be much worse: the amplifier's output noise temperature could easily be greater at the image frequency, causing the increase in system noise temperature to be even greater.

The simple solution to this problem is to use a filter in front of the mixer. The filter removes the image response, so we have, simply,

$$T_s = T_n + \frac{T_{SSB}}{G_a} \tag{9.12}$$

The filter, however, terminates the input of the mixer in a reactance, which is likely to change its noise temperature and conversion loss. Thus, the value of T_{SSB} in (9.12) is likely to differ from that in (9.11), and the noise contribution from an IF amplifier also is likely to change. To avoid unpleasant surprises, T_{SSB} and G_{cr} should be characterized with the filter in place.

Many broadband mixers have multiple responses, not just the image. For example, imagine a 1 to 1,000-MHz, doubly balanced diode mixer

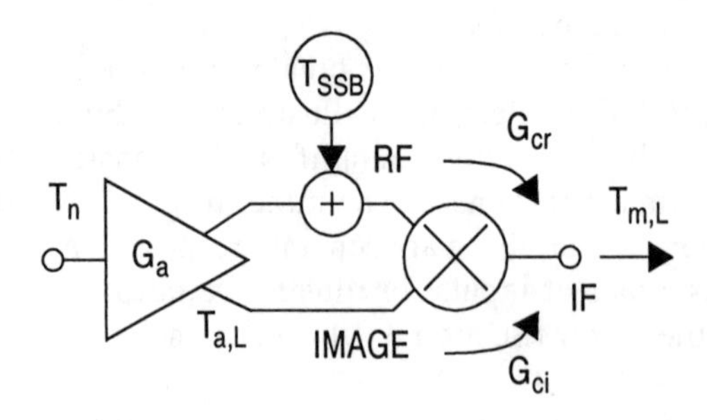

Figure 9.5 Amplifier and mixer cascade, in which the mixer has a significant image response.

operated at RF and LO frequencies near 60 MHz. Such a mixer has 16 LO harmonics within its passband! Then, virtually all significant ($\pm 1, n$) responses are terminated in the source and load impedances.

This raises yet another dilemma. In this case, the output noise of the mixer is the noise generated within the mixer, plus the downconverted noise of the terminations. Then,

$$T_{m,\,L} = T_{m,\,Ln} + \sum_{n} G_{cn} T_t \tag{9.13}$$

where $T_{m,\,Ln}$ is the mixer's output noise temperature due only to noise generated within the mixer, T_t is the termination noise temperature, and G_{cn} is the conversion loss at some nth response. These latter terms rapidly decrease with n, but it is likely that the first few are large enough to affect the mixer's, and thus the system's, noise temperature. In fact, in many broadband, low-frequency mixers, the diode switching time is fast enough, relative to a period of the LO waveform, that the diode is always a virtual open or short circuit. Then, little power is dissipated in the diode and little noise is generated by it. Most loss is caused by dissipation in the terminations at higher-order mixing frequencies. This implies, in turn, that the noise of those terminations is converted, relatively efficiently, to the IF, and may even be the dominant noise source.

How do we deal with this? The obvious solution is to use a filter, but then we encounter the same difficulty as with an image-rejection filter: the filter changes the characteristics of the mixer. Clearly, for best accuracy in the design of the system, the mixer should be characterized with the filter in place. In addition, because of the multiple responses, it becomes even more important to avoid high noise output from the LNA at the frequencies of those responses.

The specification of a mixer noise temperature is affected by this dilemma. Should the noise temperature be based simply on the internal noise of the mixer, or should it include termination noise? Depending on the intended use of the mixer, either option could be justified. Usually a broadband mixer is characterized with broadband source and load terminations. In this case, the mixer conversion loss includes the effects of those terminations, and its noise temperature includes their noise. The mixer then behaves according to the attenuator noise model, with an effective temperature close to (or perhaps slightly higher than) the temperature at which it is measured.

9.5.2 LO Noise

The LO source generates both amplitude (AM) and phase (PM) noise. Amplifiers used to increase the level of the LO signal can introduce substantial amounts of AM noise. The phase component of the amplifier noise may also increase the PM noise of the oscillator, especially at large offsets from the oscillation frequency.

If the LO's AM noise spectrum extends to the RF frequency, the noise can be downconverted to the IF through the ordinary mixing process. As such, it behaves much like noise introduced at the RF port. To prevent the degradation of the system noise temperature by LO AM noise, the LO should be well filtered at the RF and image frequencies. When this is impossible, a balanced mixer, which rejects such noise, can be employed.

9.5.3 Phase Noise

Phase noise in the local oscillator does not directly increase the noise temperature of a system using a mixer. However, a mixer is, in effect, a phase adder, so the phase noise of the LO is added, degree for degree, to the received signal. If the signal's phase carries information, as it does in a phase modulated or quadrature-amplitude modulated signal, this noise has much the same effect on demodulation as AM noise in the front end. In contrast to a balanced mixer's ability to reject AM noise, there is, unfortunately, no way to reject phase noise in a mixer.

9.6 FREQUENCY MULTIPLIERS

Frequency multipliers can be realized in a number of ways. Passive frequency multipliers use varactors, step-recovery diodes, or resistive Schottky diodes; active multipliers use FETs or bipolar devices. In all cases, the nonlinearity of a diode or transistor provides frequency multiplication. The noise characteristics of such multipliers depend largely on the type of device that is used. AM noise is rarely a concern in frequency multipliers; a greater concern is the possibility of exacerbation of phase noise.

Frequency multipliers are, in effect, phase multipliers. (Frequency is, after all, simply the time derivative of phase.) Consequently, multiplying the frequency of a signal also multiplies its phase noise by the same factor. The effect on phase noise can be derived simply from the theory of frequency modulation. Viewing the noise momentarily as a sinusoidal phase deviation, we have

$$V(t) = V_s \cos(\omega_p t + \Delta\phi \sin(\omega_m t)) \tag{9.14}$$

where $\Delta\phi$ is the peak phase deviation at the offset frequency ω_m. This can be viewed as a form of narrowband frequency modulation, with radian frequency

$$\omega = \omega_p + \Delta\phi\, \omega_m \cos(\omega_m t) \tag{9.15}$$

The peak frequency deviation is

$$\Delta\omega = \Delta\phi\, \omega_m \tag{9.16}$$

so

$$\Delta\phi = \frac{\Delta\omega}{\omega_m} \tag{9.17}$$

The theory of narrowband frequency modulation gives the magnitude of the sideband voltage, V_{SSB}:

$$\left(\frac{V_{SSB}}{V_s}\right)^2 = \mathrm{J}_1^2\left(\frac{\Delta\omega}{\omega_m}\right) = \mathrm{J}_1^2(\Delta\phi) \approx \frac{1}{4}(\Delta\phi)^2 \tag{9.18}$$

where J_1 is a first-order Bessel function. The approximation is very accurate for small phase deviations, which is certainly valid for practical levels of phase noise. It is possible to show that (9.18) is valid when $(\Delta\phi)^2$ is replaced by the mean-square phase deviation, $\overline{|\Delta\phi|^2}$.

From (9.18) it should be clear that multiplying the frequency of a signal multiplies the frequency deviation, or, viewed alternatively, the phase deviation. This results in an n^2 increase in the noise level relative to the signal, where n is the order of multiplication. Expressed in decibels, $L(f_m)$ increases as 20 log(n).

The 20 log(n) rule applies to all frequency multipliers and all frequency-multiplication phenomena. For example, it applies to harmonic mixers, in which the RF signal is mixed with a harmonic of the LO. Depending upon the type, the multiplier may introduce additional phase or amplitude noise of its own, which comes from upconversion of the device noise, especially low-frequency noise. Thus, the 20 log(n) degradation is a minimum, and a greater increase in phase noise is possible.

9.6.1 Diode Multipliers

9.6.1.1 Reactive Diodes

Varactor and step-recovery diodes use capacitive nonlinearities to generate harmonics. In such devices, $1/f$ noise, which arises only from rectified diode current, is minimal. Similarly, the reactive junction does not generate thermal noise, although a small amount is generated by the series resistance of the diode and circuit losses. These effects are almost always negligible, and the only real concern is the inevitable 20 log(n) increase in phase noise.

Figure 9.6 shows the circuit of a varactor frequency multiplier. Besides the varactor, it consists of input and output matching networks and a set of *idlers*, resonators tuned to intermediate harmonics. Idlers need not exist at all intermediate harmonics; selection of idler frequencies involves a trade-off between complexity and efficiency. The idlers are, in effect, short circuits at their respective harmonics. They are necessary for good efficiency when the order of multiplication is greater than two. The series resistance R_s is mildly nonlinear; however, it is modeled acceptably as a linear resistance. The voltage source in series with R_s models its thermal noise.

Step-recovery diode multipliers operate somewhat differently from conventional varactors. The step-recovery diode is capable of large diffusion charge storage when forward biased. When the device is driven

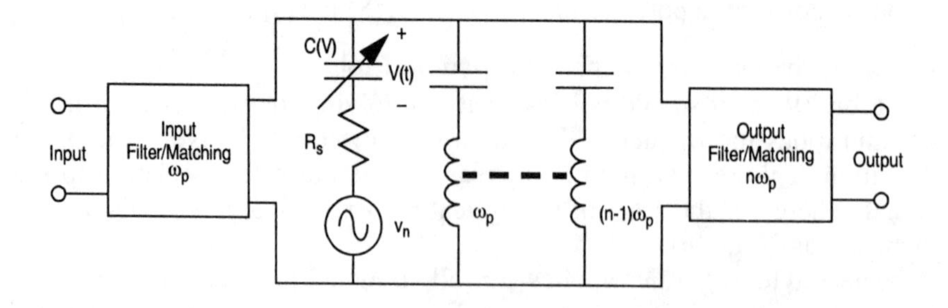

Figure 9.6 Varactor frequency multiplier, consisting of input and output matching circuits, the varactor, and a set of idlers at a number of intermediate harmonics.

through an inductive source, diffusion charge is completely removed at the peak of the inductor's current. The diode then switches suddenly to a low-capacitance mode. At that point, the diode and the inductor realize a series LC resonator, excited by the inductor's current. The result is oscillation, at the LC resonant frequency. The oscillation only lasts one-half cycle, however, as the diode then turns on and stops the oscillation. That half-cycle, is, in effect, a large, fast voltage pulse having substantial harmonic content.

Reactive diodes are theoretically capable of lossless multiplication. In practice, well-designed varactor multipliers exhibit a few decibels of loss, perhaps 3 dB at microwave frequencies for a doubler. They are relatively narrowband. Step-recovery diodes are used to generate high harmonics, pulses, or as comb generators, which produce a wide spectrum of harmonics, approximately of equal amplitude.

The strong capacitive nonlinearities of reactive-diode multipliers make them subject to various types of instability. Multiplier instabilities are fundamentally chaotic, so it is difficult to make generalizations about them. The instabilities can take many forms; one common one is the generation of low-frequency components that appear as multiple sidebands around the excitation frequency and its harmonics. Often, the oscillation, being chaotic, is noise-like and can be mistaken for a simple noise process. Another manifestation of instability is the exacerbation of noise sidebands of the excitation signal; this also can be mistaken for noise generated within the multiplier.

9.6.1.2 Resistive Diodes

Schottky-barrier diodes are also used as multipliers. Schottky diodes are resistive devices, and their junction current generates shot and $1/f$ noise. Schottky-diode multipliers are lossy. The theoretical maximum efficiency of a resistive multiplier is $1/n^2$, where n is the order of multiplication. Resistive frequency doublers usually have conversion losses of approximately 10 dB.

Because of their high loss, multipliers having $n > 2$ are rarely practical. In compensation for their loss, however, resistive multipliers are capable of wide, often multioctave, bandwidths, much wider than reactive multipliers. Instability of Schottky-diode multipliers, caused by pumped junction capacitance, is possible although rare. The loading of the junction capacitance by the average value of the junction conductance usually is enough to prevent reactive instability.

The pumping of the diode junction creates a mixing process that upconverts the diode's $1/f$ noise to the output frequency. The diode also generates high-frequency shot noise, which has a component at the output frequency, and is also up- and downconverted from all other mixing frequencies. The conversion of shot noise in a resistive multiplier is fundamentally the same as the conversion of such noise in mixers. The phase component of the diode's noise can introduce additional phase noise into the signal. These are small effects, however, and except in very low-noise applications, the diode noise is usually negligible.

9.6.2 Active Multipliers

Both FETs and bipolar transistors have been used as active multipliers. Such multipliers are capable of moderate output power (tens of milliwatts with ordinary, small-signal devices) and good conversion gain (approximately unity). This makes them ideal for frequency multipliers in receiver LO chains and similar applications.

As an active multiplier, the FET or bipolar transistor is biased at the gate or base near the turn-on point of the device. When the device is driven by a sinusoidal signal, it conducts in pulses. The duty cycle of the drain or collector current can be adjusted by the bias voltage to optimize the frequency component at the desired harmonic. FET frequency multipliers are practical for doublers and triplers, but the difficulty of generating short, high-amplitude pulses of drain current, which are necessary for high-order multiplication, make them unsuitable for generating higher harmonics. Bipolars are somewhat better for high-harmonic multipliers, but still are limited to a factor of three or four. Above this, the gain and output power do not compete well with diodes.

The rules for optimizing active multipliers are generally the same as those for active mixers: the practical optimum is a short-circuit termination, at the input and output, at all unwanted harmonics. The short-circuit termination is especially important for the fundamental-frequency component at the drain; if this is not well terminated, instability results. It is possible to increase conversion efficiency by modifying this termination, but only at the expense of decreased stability margin [9.7]. When it is difficult to terminate the drain/collector of a single device, a balanced multiplier can be used. The balanced multiplier provides an optimum termination without the use of filters.

Because of their relatively high low-frequency gain and large junction capacitance, bipolars are somewhat more prone to instability than FETs. Therefore, obeying the rules regarding termination of the device at unwanted harmonics is especially important. As with diode multipliers, instability in bipolar or FET multipliers can look like noise or can exacerbate the noise generation in the device.

Microwave FETs have relatively high levels of $1/f$ noise, which can be a source of increased phase noise in active multipliers. Bipolar devices have lower $1/f$ noise, but operate at lower frequencies, so they may not be useful in the upper end of the microwave range. HBTs may be a reasonable option at higher frequencies, where homojunction bipolars are not adequate.

Active multipliers can also generate amplitude noise. The AM noise of a multiplier is not likely to be significantly greater than that of an amplifier, so much the same caveats apply to multiplier AM noise as to amplifier AM noise (Section 9.5.2).

References

[9.1] D. N. Held and A. R. Kerr, "Conversion Loss and Noise of Microwave and Millimeter-Wave Mixers: Part 1—Theory," *IEEE Trans. Microwave Theory Tech.*, Vol. MTT-26, no. 3, p. 49, 1978.

[9.2] S. A. Maas, "A GaAs MESFET Mixer with Very Low Intermodulation," *IEEE Trans. Microwave Theory Tech.*, Vol. MTT-35 p. 425, 1987.

[9.3] S. Maas, "A GaAs MESFET Balanced Mixer with Very Low Intermodulation," *IEEE MTT-S International Microwave Symposium Digest*, p. 895, 1987.

[9.4] C. Dragone, "Analysis of Thermal Shot Noise in Pumped Resistive Diodes," *Bell Syst. Tech. J.*, Vol. 47, p. 1883, 1968.

[9.5] S. Maas, "Two-Tone Intermodulation in Diode Mixers," *IEEE Trans. Microwave Theory Tech.*, Vol. MTT-35, no. 3, p. 307, 1987.

[9.6] S. Maas, *Microwave Mixers*, 2nd ed., Norwood, MA: Artech House, 1993.

[9.7] C. Rauscher, "High-Frequency Doubler Operation of GaAs Field-Effect Transistors," *IEEE Trans. Microwave Theory Tech.*, Vol. MTT-31, p. 462, 1983.

About the Author

Stephen A. Maas received BSEE and MSEE degrees in electrical engineering from the University of Pennsylvania in 1971 and 1972, respectively, and a Ph.D. in electrical engineering from UCLA in 1984. He joined the National Radio Astronomy Observatory in 1974, where he designed the low-noise receivers for the Very Large Array radio telescope. Subsequently, at Hughes Aircraft Company and TRW, he developed low-noise microwave and millimeter-wave systems and components, primarily FET amplifiers and diode and FET mixers, for space communication. He also has been employed as a research scientist at The Aerospace Corporation, where he worked on the optimization of nonlinear microwave circuits and the development of circuit-design software based on harmonic-balance, Volterra-series, and time-domain methods. He joined the UCLA Electrical Engineering Faculty in 1990 and left it in 1992. Since then, he has worked as an independent consultant and currently is the chief scientist at Applied Wave Research, Inc.

Dr. Maas is the author of three other books, *Microwave Mixers* (1986 and 1992), *Nonlinear Microwave Circuits* (1988 and 2003), and *The RF and Microwave Circuit Design Cookbook* (1998), all published by Artech House. From 1990 until 1992 he was the editor of the *IEEE Transactions on Microwave Theory and Techniques*, and from 1990 to 1993 he was a member of the MTT Administrative Committee and publications chairman of the IEEE MTT Society. He received the MTT Society's Microwave Prize in 1989 for his work on distortion in diode mixers and its Application Award in 2002 for his invention of the FET resistive mixer. He is a fellow of the IEEE.

About the Author

Index

D

E

F

G

H

O

P

Q

R

S

Recent Titles in the Artech House Microwave Library

Active Filters for Integrated-Circuit Applications, Fred H. Irons

Advanced Techniques in RF Power Amplifier Design, Steve C. Cripps

Automated Smith Chart, Version 4.0: Software and User's Manual, Leonard M. Schwab

Behavioral Modeling of Nonlinear RF and Microwave Devices, Thomas R. Turlington

Broadband Microwave Amplifiers, Bal S. Virdee, Avtar S. Virdee, and Ben Y. Banyamin

Computer-Aided Analysis of Nonlinear Microwave Circuits, Paulo J. C. Rodrigues

Design of FET Frequency Multipliers and Harmonic Oscillators, Edmar Camargo

Design of Linear RF Outphasing Power Amplifiers, Xuejun Zhang, Lawrence E. Larson, and Peter M. Asbeck

Design of RF and Microwave Amplifiers and Oscillators, Pieter L. D. Abrie

Digital Filter Design Solutions, Jolyon M. De Freitas

Distortion in RF Power Amplifiers, Joel Vuolevi and Timo Rahkonen

EMPLAN: Electromagnetic Analysis of Printed Structures in Planarly Layered Media, Software and User's Manual, Noyan Kinayman and M. I. Aksun

FAST: Fast Amplifier Synthesis Tool—Software and User's Guide, Dale D. Henkes

Feedforward Linear Power Amplifiers, Nick Pothecary

Fundamentals of Nonlinear Behavioral Modeling for RF and Microwave Design, John Wood and David E. Root, editors

Generalized Filter Design by Computer Optimization, Djuradj Budimir

High-Linearity RF Amplifier Design, Peter B. Kenington

High-Speed Circuit Board Signal Integrity, Stephen C. Thierauf

Intermodulation Distortion in Microwave and Wireless Circuits, José Carlos Pedro and Nuno Borges Carvalho

Lumped Elements for RF and Microwave Circuits, Inder Bahl

Microwave Circuit Modeling Using Electromagnetic Field Simulation, Daniel G. Swanson, Jr. and Wolfgang J. R. Hoefer

Microwave Component Mechanics, Harri Eskelinen and Pekka Eskelinen

Microwave Engineers' Handbook, Two Volumes, Theodore Saad, editor

Microwave Filters, Impedance-Matching Networks, and Coupling Structures, George L. Matthaei, Leo Young, and E.M.T. Jones

Microwave Materials and Fabrication Techniques, Second Edition, Thomas S. Laverghetta

Microwave Mixers, Second Edition, Stephen A. Maas

Microwave Radio Transmission Design Guide, Trevor Manning

Microwaves and Wireless Simplified, Third Edition, Thomas S. Laverghetta

Modern Microwave Circuits, Noyan Kinayman and M. I. Aksun

Modern Microwave Measurements and Techniques, Second Edition, Thomas S. Laverghetta

Neural Networks for RF and Microwave Design, Q. J. Zhang and K. C. Gupta

Noise in Linear and Nonlinear Circuits, Stephen A. Maas

Nonlinear Microwave and RF Circuits, Second Edition, Stephen A. Maas

QMATCH: Lumped-Element Impedance Matching, Software and User's Guide, Pieter L. D. Abrie

Practical Analog and Digital Filter Design, Les Thede

Practical RF Circuit Design for Modern Wireless Systems, Volume I: Passive Circuits and Systems, Les Besser and Rowan Gilmore

Practical RF Circuit Design for Modern Wireless Systems, Volume II: Active Circuits and Systems, Rowan Gilmore and Les Besser

Production Testing of RF and System-on-a-Chip Devices for Wireless Communications, Keith B. Schaub and Joe Kelly

Radio Frequency Integrated Circuit Design, John Rogers and Calvin Plett

RF Design Guide: Systems, Circuits, and Equations, Peter Vizmuller

RF Measurements of Die and Packages, Scott A. Wartenberg

The RF and Microwave Circuit Design Handbook, Stephen A. Maas

RF and Microwave Coupled-Line Circuits, Rajesh Mongia, Inder Bahl, and Prakash Bhartia

RF and Microwave Oscillator Design, Michal Odyniec, editor

RF Power Amplifiers for Wireless Communications, Steve C. Cripps

RF Systems, Components, and Circuits Handbook, Ferril Losee

Stability Analysis of Nonlinear Microwave Circuits, Almudena Suárez and Raymond Quéré

TRAVIS 2.0: Transmission Line Visualization Software and User's Guide, Version 2.0, Robert G. Kaires and Barton T. Hickman

Understanding Microwave Heating Cavities, Tse V. Chow Ting Chan and Howard C. Reader

For further information on these and other Artech House titles, including previously considered out-of-print books now available through our In-Print-Forever® (IPF®) program, contact:

Artech House Publishers
685 Canton Street
Norwood, MA 02062
Phone: 781-769-9750
Fax: 781-769-6334
e-mail: artech@artechhouse.com

Artech House Books
46 Gillingham Street
London SW1V 1AH UK
Phone: +44 (0)20 7596 8750
Fax: +44 (0)20 7630 0166
e-mail: artech-uk@artechhouse.com

Find us on the World Wide Web at: www.artechhouse.com